THE STUDY OF ENZYME MECHANISMS

THE STUDY OF
ENZYME MECHANISMS

EUGENE ZEFFREN
Miami Valley Laboratories
The Procter and Gamble Company

PHILIP L. HALL
Department of Chemistry
Virginia Polytechnic Institute and State University

A WILEY-INTERSCIENCE PUBLICATION

JOHN WILEY & SONS, New York · London · Sydney · Toronto

Copyright © 1973, by John Wiley & Sons, Inc.

All rights reserved. Published simultaneously in Canada.

No part of this book may be reproduced by any means, nor transmitted, nor translated into a machine language without the written permission of the publisher.

Library of Congress Cataloging in Publication Data

Zeffren, Eugene, 1941–
The study of enzyme mechanisms.

"A Wiley-Interscience publication."
Includes bibliographical references.
1. Enzymes. I. Hall, Philip L., joint author.
II. Title.

QP601.Z37   574.1'925   72–13751
ISBN 0–471–98150–8

Printed in the United States of America
10-9 8 7 6 5 4 3 2 1

Preface

The amount of research being carried out in laboratories around the world to elucidate the mechanism of enzyme action has virtually exploded since 1960. These studies have gone much further than just the definition of the stoichiometry of a reaction, and indeed, in the cases of some enzymes, the characterization of some short-lived intermediates is now being attempted.

Certainly these developments have not escaped the attention of textbook authors. Virtually all biological chemistry texts treat this area to some extent although the coverage is necessarily limited and selective. On the other hand a number of advanced texts, monographs, and reviews are now available covering various topics relevant to the study of enzyme mechanisms in detail. The writing of this book was undertaken to fill the gap that exists between the broad general treatment in the standard biochemistry textbooks and the deeper but more specific treatment of the advanced monographs and reviews. In addition, we hope that this book will now make it easier to bring the fascination of enzyme mechanism studies to the attention of upper-class undergraduates and beginning graduate students, many of whom are receiving their first exposure to research at this time.

This book is divided into two major sections. Part One provides the theoretical and practical framework required to carry out meaningful studies into the mechanism of enzyme action. Thus we have chapters on enzyme methods and chemical modification studies as well as basic enzyme kinetics, relevant organic mechanisms, and coenzyme involvement. The treatment provided will allow the researcher to attain a significant level of understanding of enzyme mechanism.

Part Two reviews selected areas of enzymology to illustrate the principles discussed in the first section. The book concludes with a chapter on allosterism, a concept of exceeding importance to the understanding of the regulation of biological reactions in living systems.

If we have been successful, this volume will give the student a modest understanding and appreciation of current research in the field of enzyme mechanisms.

For students who seek further sophistication in this area, this book will give a broader foundation than is attainable from standard biochemistry textbooks.

At the same time, because of the fundamental importance of enzymes to all life processes and living organisms, including animals, plants, and microorganisms, we feel that this material will be of interest and value to those students not going on to graduate work in chemistry and biochemistry. Virtually every reaction that occurs within the human body, save the reaction of hydronium and hydroxide ions, is mediated by some type of enzyme. Among the diseases of man know to have their causes in the malfunctioning of an enzyme or enzyme system are phenylketonuria and Parkinson's disease. Indeed, some enzymes have found therapeutic use, with the most spectacular example being the use of the enzyme L-asparaginase to combat some forms of leukemia.

It is only within the past decade that industry has been able to make the powerful catalytic effects of enzymes universally available to the general public. It is quite likely that in addition to their use in household detergents, further examples of large-scale usage of enzymes will develop.

With these aims, we view this book as a companion text in advanced undergraduate or beginning graduate courses in biochemistry and organic chemistry or as a text for a short course in enzyme mechanisms. Our treatment assumes that the student has completed a standard undergraduate course in organic chemistry and has had some introduction to physical chemistry. No formal training in biochemistry is assumed.

Acknowledgments

We are grateful to our wives, Barbara Taylor Hall and Steccia Leigh Zeffren, for their help in the preparation of the manuscript and illustrations. We are deeply appreciative of the effort of Prof. E. T. Kaiser for a critical review of the manuscript and for his encouragement along the way. We are also grateful to Pharmacia Fine Chemicals Corp. and to LKB Instruments, Inc., for the use of photographic materials.

EUGENE ZEFFREN
PHILIP L. HALL

Cincinnati, Ohio
Blacksburg, Virginia
November, 1972

Contents

THE STUDY OF ENZYME MECHANISMS

PART ONE

General Principles

CHAPTER 1

Introduction

General Description of Proteins

It is universally accepted that all enzymes are proteins, although some have crucial dependencies on nonprotein prosthetic groups, or cofactors. Although this seems axiomatic now, it was not until 1926 that an enzyme was obtained in crystalline form by Sumner[1] (the enzyme was a urease isolated from jack bean, which catalyzes the cleavage of urea to ammonia and carbon dioxide), and it took almost a decade of work by Northrop[2] and his collaborators on a number of enzymes to convince the more skeptical researchers of the day of the protein nature of enzymes. Since enzymes are indeed proteins, one cannot hope to begin explaining enzyme catalysis without first providing a general framework for the description of proteins.

In general, proteins are made of many amino acids of the L-configuration linked by an amide bond between the carboxyl group of one amino acid and the α-amino of another—the peptide bond. The occurrence of D amino acids in normal mammalian proteins is extremely rare. In this book, all amino acid residues will be understood to be of the L-configuration unless otherwise specified.

A symbolic representation of a protein in aqueous solution near neutrality, then, is shown in structure **I**, where n is any integer greater than about 75*

$$\overset{\oplus}{H_3N}-\underset{\underset{\textbf{I}}{}}{\overset{\overset{R}{|}}{CH}}-\left[\overset{\overset{O}{\|}}{C}-NH-\overset{\overset{R}{|}}{CH}\right]_n\overset{\overset{O}{|}}{C}\overset{\ominus}{-}O$$

and R may be any one of about 20 or more common amino acid side chains. Although many enzymes are considered to be polypeptides, a polypeptide is usually not considered to be a protein unless n is minimally from about 40 to 75. This lower limit is rather arbitrary and not well agreed to by various authors. By convention, protein structural representation starts with the

---

* Lysozyme is one of the smallest common enzymes, containing just over 100 amino acid residues, as discussed in Chapter 9.

amino terminus (N-terminus) at the left of the page and works to the right, with the last residue being termed the carboxyl terminus (C-terminus). Thus the tripeptide aspartyl-phenylalanyl-serine has the structure **II**. Table

$$
\overset{\oplus}{H_3}N-\underset{\underset{\underset{COOH}{|}}{\overset{|}{CH_2}}}{CH}-\overset{\overset{O}{\|}}{C}-NH-\underset{\underset{\underset{\bigcirc}{|}}{\overset{|}{CH_2}}}{CH}-\overset{\overset{O}{\|}}{C}-NH-\underset{\underset{\underset{OH}{|}}{\overset{|}{CH_2}}}{CH}-COO^{\ominus}
$$

II

1-1 lists the names, structures, and abbreviations for the common amino acids. The side chains of most of these compounds are chemically inert. Those amino acids that have proven to be important in catalytic enzymology are serine, histidine, aspartic acid, glutamic acid, cysteine, tyrosine, lysine, and arginine.

The sequential arrangement of L amino acids in a protein molecule is commonly referred to as the protein's *primary structure*. Since the bonds forming this chain are formally single bonds,* one might think that a molecule composed of 100 or more α-amino acids linked as in structure **II** would be able to assume an infinite number of three-dimensional orientations in solution, and in principle it can. However, when a protein is in its native state, that is, the conformation of the molecule when it is biologically active, there is generally one preferred orientation of all the atoms, and this is often referred to as the *folded* (native) *state*. Certain experimental conditions (e.g., heat or extremes of pH) may cause the protein to unfold to the *denatured state*.†

The folded state derives much of its stability from two sets of noncovalent interactions, whose nature is actually determined by the primary structure of a protein. The first of these two interactions is the hydrogen bonding interaction, which takes place mainly between the proton on the α-amino nitrogen atom in the peptide bond and the oxygen atom of a carbonyl

---

* Actually, the peptide bond has a fair amount of double bond character since structures of the type $R-\underset{\underset{O^{\ominus}}{|}}{C}=\overset{\oplus}{N}H-R'$ contribute significantly to the overall structure of a protein in aqueous solution.

† In this text a denatured protein is considered to be one that has lost its biological activity in a process that is theoretically, at least, reversible, and where no covalent bonds have been either cleaved or synthesized. If such covalent changes occur and lead to loss of activity, the protein thus derived will be considered to have been *inactivated*.

TABLE 1-1 Most Common Naturally Occurring α-Amino Acids

Class	Amino Acid	Abbreviation	Molecular Weight	Structure
I. Aliphatic A. Hydrophilic	1. Glycine	Gly	75.1	H_2N-CH_2-COOH
	2. Alanine	Ala	89.1	$CH_3-\underset{\underset{NH_2}{\mid}}{CH}-COOH$
	3. Serine	Ser	105.1	$HO-CH_2-\underset{\underset{NH_2}{\mid}}{CH}-COOH$
	4. Threonine	Thr	119.1	$CH_3-\underset{\underset{OH}{\mid}}{CH}-\underset{\underset{NH_2}{\mid}}{CH}-COOH$
B. Hydrophilic-dibasic	5. Lysine	Lys	146.2	$\underset{\underset{NH_2}{\mid}}{CH_2}CH_2-CH_2-CH_2-\underset{\underset{NH_2}{\mid}}{CH}-COOH$
	6. Arginine	Arg	174.2	$H_2N-\underset{\underset{NH}{\parallel}}{C}-\underset{\underset{NH}{\mid}}{}CH_2-CH_2-CH_2-\underset{\underset{NH_2}{\mid}}{CH}-COOH$
C. Dicarboxylic acids and amides	7. Aspartic acid	Asp	133.1	$HOOC-CH_2-\underset{\underset{NH_2}{\mid}}{CH}-COOH$
	8. Asparagine	Asn	132.1	$H_2N-\underset{\underset{O}{\parallel}}{C}-CH_2-\underset{\underset{NH_2}{\mid}}{CH}-COOH$
	9. Glutamic acid	Glu	147.1	$HOOC-CH_2-CH_2-\underset{\underset{NH_2}{\mid}}{CH}-COOH$

(continued)

5

TABLE 1-1 *(continued)*

Class	Amino Acid	Abbreviation	Molecular Weight	Structure
C. Dicarboxylic acids and amides *(cont.)*	10. Glutamine	Gln	146.1	$H_2N{-}\underset{\underset{O}{\parallel}}{C}{-}CH_2{-}CH_2{-}\underset{\underset{NH_2}{\mid}}{CH}{-}COOH$
D. Hydrophobic	11. Leucine	Leu	131.2	$CH_3{-}\underset{\underset{CH_3}{\mid}}{CH}{-}CH_2{-}\underset{\underset{NH_2}{\mid}}{CH}{-}COOH$
	12. Isoleucine	Ile	131.2	$CH_3{-}CH_2{-}\underset{\underset{CH_3}{\mid}}{CH}{-}\underset{\underset{NH_2}{\mid}}{CH}{-}COOH$
	13. Valine	Val	117.1	$CH_3{-}\underset{\underset{CH_3}{\mid}}{CH}{-}\underset{\underset{NH_2}{\mid}}{CH}{-}COOH$
E. Sulfur-containing amino acids	14. Cysteine	Cys	121.2	$HS{-}CH_2{-}\underset{\underset{NH_2}{\mid}}{CH}{-}COOH$
	15. Cystine	Cys[a]	240.3	$S{-}CH_2{-}\underset{\underset{NH_2}{\mid}}{CH}{-}COOH$ $\mid$ $S{-}CH_2{-}\underset{\underset{NH_2}{\mid}}{CH}{-}COOH$
	16. Methionine	Met	149.2	$CH_3{-}S{-}CH_2{-}CH_2{-}\underset{\underset{NH_2}{\mid}}{CH}{-}COOH$

F. Heterocyclic

17. Proline · Pro · 115.1

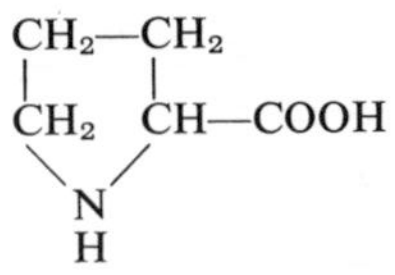

$$CH_2-CH_2$$
$$CH_2-CH-COOH$$ (ring closed through N–H)

II. Aromatic
A. Hydrophobic

18. Phenylalanine · Phe · 165.2

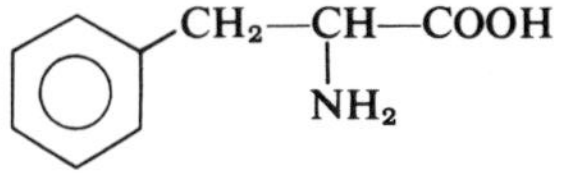

$$CH_2-CH-COOH$$
$$|$$
$$NH_2$$
(phenyl ring)

19. Tyrosine · Tyr · 181.2

$$CH_2-CH-COOH$$
$$|$$
$$NH_2$$
(HO– phenyl ring)

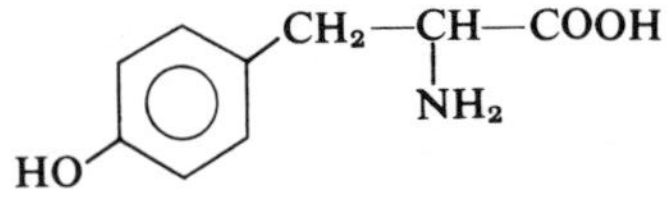

B. Hydrophobic-heterocyclic

20. Tryptophan · Trp · 204.2

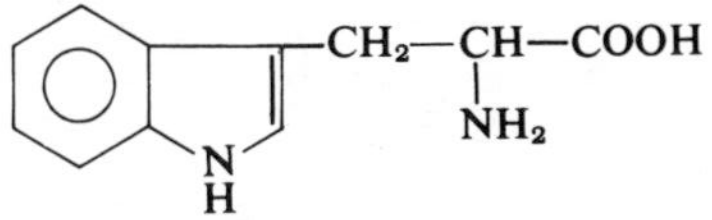

$$CH_2-CH-COOH$$
$$|$$
$$NH_2$$
(indole ring, N–H)

C. Hydrophilic-heterocyclic-dibasic

21. Histidine · His · 155.2

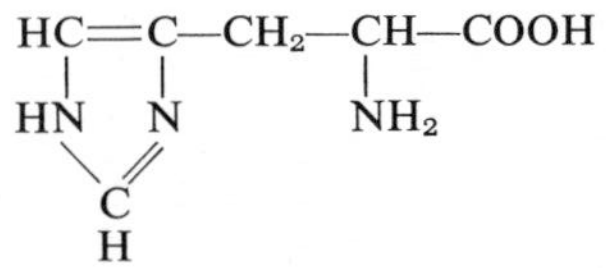

$$HC=C-CH_2-CH-COOH$$
$$|$$
$$NH_2$$
$$HN \quad N$$
$$\diagdown C \diagup$$
$$H$$

---

[a] Cystine is seen from the structure to be a dimer of cysteine. In protein sequence studies it has been found that the two amino groups and two carboxyl groups of cystine are always involved in the primary sequence in peptide linkages. Thus in published sequence studies, cystine is usually abbreviated as $\frac{1}{2}$cystine, or $\frac{1}{2}$Cys. These authors have occasionally found it useful in discussing the organic chemistry of cysteine and cystine to refer to them as CysH and CysS, respectively.

carbon. There may be important hydrogen bonding between side chain proton donors and acceptors as well, and this is the case with many enzymes.

In general, one H-bond of the type $R\!-\!C\!\!=\!\!O\cdots H\!-\!N{\Large<}$ will contribute about 800 cal/mole to the stabilization of a protein's structure.* In the protein structural form called the α-helix the carbonyl oxygen atom of 1 residue is hydrogen bonded to a proton on an amino nitrogen atom 3 residues away, and it requires roughly $3\frac{1}{2}$ residues per turn of the helix (see Figure 1-1).

The α-helix is only one form of hydrogen-bonded structure that may form in proteins; many enzymes are globular proteins with less than 30% of the molecule in the α-helical state. For example, the proteolytic enzyme pepsin has been estimated to have less than 10% helical structure in its native state. Another common hydrogen-bonded structure has been referred to as the "pleated sheet" or the β-structure, and this is also important in determining the overall shape of a protein molecule. X-ray crystallography has vividly shown this to be the case for the enzyme carboxypeptidase. This hydrogen-bonded structure, whether α-helical, pleated sheet, or random in nature is referred to as a protein's *secondary structure*.[4]

When a protein has had its primary and secondary structure established, the other set of noncovalent interactions mentioned earlier come into play. These are van der Waals attractions and "hydrophobic" bonds,† and these interactions determine the relative orientation of the side chains of the various amino acids in a protein. Since the functional groups of some side chains are critically involved in an enzyme's catalytic process, these weak interactions (along with hydrogen-bonding interactions) are quite important in enzymology. Typically, the more polar side chains, like those of lysine, arginine, glutamic acid, and aspartic acid, are found on the surface of the molecule, whereas the less polar, hydrocarbonlike side chains, such as those of valine, leucine, phenylalanine, and tryptophan, are located in the interior of the molecule. This is a generalization, of course, and there are some important exceptions in some enzymes. One exception is the ion pair that has been shown to exist in chymotrypsin in a hydrophobic pocket, and the disruption of this ion pair at pH values above about 9.0 leads to the enzyme's losing its proteolytic

---

* This is inferred from the estimate of Jencks[3] for the contribution of hydrogen bonding to the overall binding of *tri*-N-acetylglucosamine to lysozyme. It is assumed that intra-molecular H-bonding within the enzyme has the same energy content as H-bonding between enzyme and substrate.

† A hydrophobic bond in a protein may be thought of as an attractive interaction between the apolar side chains of the protein's constituent amino acids. This interaction reflects the preference of these hydrocarbonlike groups for a medium of lower dielectric constant than that of water.[5]

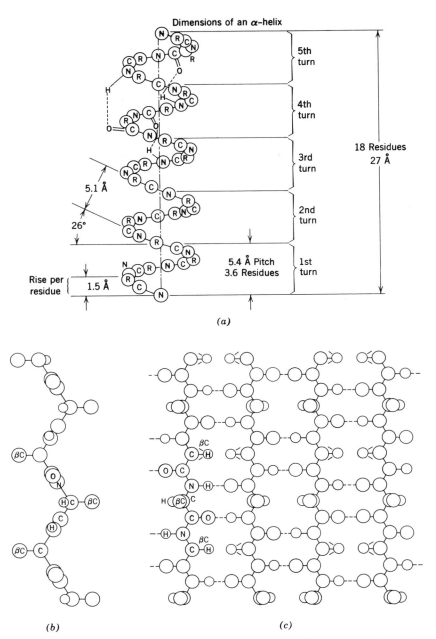

FIGURE 1-1. (*a*) Diagram of an α-helix indicating the 1.5-Å rise per residue and the relation between the pitch (5.4 Å; 3.6 residues) and the 5.1-Å spacing. Also indicated are a few of the many H-bonds stabilizing the structure. (*b*) Diagram of the pleated sheet protein structure viewed parallel to the planes of the sheet. (*c*) Diagram of the pleated sheet protein structure viewed perpendicular to the sheet. (After E. H. Mercer, *Keratins and Keratinization*, Pergamon Press, New York, 1961, pp. 179, 182.)

9

activity. Free energies of hydrophobic interactions are roughly equivalent to those of hydrogen bonding; they determine the protein's *tertiary structure*, and in terms of bond strengths they also measure about 800 cal/mole/interaction.

Some proteins exist as aggregates of a small number of identical subunits (from 2 to 6) or as aggregates of sets of subunits. Hemoglobin is such a molecule, consisting of four subunits—two pairs of two identical subunits. Each subunit is a complete protein itself, but the biological function of the hemoglobin molecule requires the proper interrelationship of all four subunits. The interactions that are responsible for holding the subunits together are of the hydrogen-bonding and van der Waals noncovalent type, and they determine what is referred to as a protein's *quaternary structure*. A detailed treatment of such complex systems is beyond the scope of this book, but they are given some further consideration in Chapter 12 since they are extremely important systems biologically.

Thus we see the overall nature of the catalysts that concern us: nonaggregated polypeptide materials of high molecular weight that catalyze organic (and inorganic) reactions. The field of coenzymes will be treated in Chapter 8, where a selection of important coenzymes will be examined with respect to what is known about their function. This is actually a large field of research in its own right and is worthy of a separate text.

General Enzymatic Phenomena

It is quite astounding, actually, that a material composed solely of amino acids can so markedly affect the rate of an organic reaction. For example, at a concentration of about $10^{-6}M$, the enzyme pepsin will completely hydrolyze the dipeptide N-acetyl-L-phenylalanyl-L-tyrosine in one hour when the dipeptide is present at an initial concentration of $10^{-3}M$ in a pH 2.0 buffer at 35°C. Yet if the enzyme is omitted from the reaction mixture, the dipeptide in solution will remain unhydrolyzed almost indefinitely—certainly for a period of months. This remarkable feat is not unusual for enzymatic reactions, and a large measure of their fascination for the chemist is this extraordinary efficiency. When the rates of corresponding nonenzymatic reactions can be measured conveniently, the enzyme system is usually found to be more favorable by from two to eight *orders of magnitude*, with factors of from four to five orders of magnitude rate increase being quite common.

This overall catalytic efficiency usually is attributed to a combination of three general aspects of the enzymic process: (1) the formation of a noncovalent complex with the substrate, or simply, binding; (2) the manifestation of substrate "specificity" by the enzyme, that is, the relative preference of an enzyme for one substrate over another (stated in terms of another interpretation, specificity consists of an enzyme's ability to discern between optical isomers of a material, cleaving one but not another); and (3) the catalytic process itself.

The binding phenomenon is usually considered to be an equilibrium process resulting in the formation of the noncovalent ES complex—the Michaelis complex—as shown in Eq. 1-1.

$$E + S \underset{k_{-1}}{\overset{k_1}{\rightleftharpoons}} ES \qquad K_s = \frac{k_{-1}}{k_1} = \frac{(E)(S)}{(ES)} \qquad (1\text{-}1)$$

It is characterized by a dissociation constant K_s for the reverse reaction, the dissociation of ES into E and S.*

Many compounds may bind to an enzyme, but only certain types will lead to subsequent reaction; these are called *substrates* and they satisfy the particular enzyme's specificity requirements. Materials that bind but do not thereupon chemically react often affect the overall rate of an enzymic reaction in either a positive or negative way because the presence of a bound, unreactive species can alter the catalytic properties of an enzyme. Compounds of the former class are called *activators*, those of the latter are *inhibitors*; generally, inhibition is more common than activation. The inhibition process is also described by a dissociation constant, K_i, the inhibition constant (Eq. 1-2).

$$E + I \underset{k_{-1'}}{\overset{k_{1'}}{\rightleftharpoons}} EI; \qquad K_i = \frac{k_{-1'}}{k_{1'}} = \frac{(E)(I)}{(EI)} \qquad (1\text{-}2)$$

Many enzymes exhibit strict specificity, reacting with only one substrate. A common type of specificity is *configurational stereospecificity*; here the enzyme hydrolyzes only one of an enantiomeric pair of compounds or only one of a diastereomeric set of compounds. Thus, in the pepsin system mentioned earlier, N-acetyl-L-phenylalanyl-L-tyrosine is readily cleaved to N-Ac-L-Phe and L-Tyr. The replacement of either amino acid residue by its D enantiomer or replacement of both leads to three different diastereomers of N-acetylphenylalanyl-tyrosine, the L-D, D-L, and D-D, none of which are cleaved by pepsin. On the other hand, all of these molecules are inhibitors of pepsin and retard its rate of action. Hydrolytic enzymes exhibiting optical specificity show predominantly L specificity, with the exceptions usually being nonordinary types of substrates.

The molecular mechanisms of binding and exhibition of specificity are very poorly understood, although it is recognized that the enzyme's three-

* The Michaelis constant, K_m, as classically derived is sometimes equivalent to K_s. However, more detailed treatments of enzyme kinetics have revealed K_m to be a more complex quantity, in general, than K_s. (See Chapter 4.)

dimensional structure in solution determines both of these processes through the regulation of van der Waals, charge-transfer, and hydrogen-bond contacts. This book is concerned primarily with the third general characteristic, the catalytic process. This too is highly dependent upon the enzyme's maintenance of the proper conformation in solution, since this conformation will determine the interaction distances between the enzyme's catalytic functional groups and the substrate's susceptible bond.

The Active Site[6]

As a major outgrowth of the study of enzyme-catalyzed reactions, enzymologists have developed the so-called active-site hypothesis for the specific interaction of enzymes with their substrates. Generally, a substrate molecule is small relative to the enzyme molecule itself, and even for a very large substrate molecule, it is only a very small portion of the molecule that undergoes *reaction* under the influence of the enzyme. So in attempting to characterize the interaction of enzyme with substrate it is useful (and no doubt essentially correct) to visualize a specific, localized, and relatively small three-dimensional region on or near the enzyme surface within which all enzyme-substrate interactions take place. This region is referred to as the enzyme's *active site*. It encompasses all the amino acid side chains of the protein which individually and in concert participate in binding the substrate to the enzyme and/or carrying out the subsequent catalytic event itself.

The amino acids belonging to the active site need not be close together in the primary sequence of the enzyme, since the folding of the native structure can bring groups well separated in the amino acid sequence into intimate contact with one another. For example, two crucial catalytic groups in the proteolytic enzyme chymotrypsin are histidine residue 57 and serine residue 195, a separation of 138 residues. X-ray crystallographic determination of the three-dimensional structure of this enzyme has shown that these two residues may be hydrogen bonded to one another in the native structure and thus can be no more than 2 to 3 Å (Ångstroms) apart. Chemical and physical alterations of amino acid residues in or near the active site of many enzymes have contributed much to an understanding of their mechanism of action and many of these changes will be considered in subsequent chapters. The binding process just discussed generally occurs at the active site, and the enzymes we are concerned with have only one active site per molecule.

The initial concept of an active site may be attributed to Emil Fisher, whose early proposal of a "lock and key" theory of enzyme specificity first put forward in the 1890s is still appealing in some respects. In this theory the enzyme's active site is conceived of as a rigid structure with a well-defined geometry, similar to a lock; it may be fitted (will react with) only by those molecules with a complementary geometry, that is, a key. More recently, researchers such as Koshland[6] and Gutfreund[7] have presented conceptions of enzymic catalysis as a more dynamic process involving cooperative interactions between an enzyme and its substrate. In these theories the active site is still thought to be localized in space, but it is more flexible. This concept is intuitively more attractive since it is unlikely that a molecule as large as an enzyme would have a region of about 125 Å^3 that is completely rigid. Whether the ultimate conception of an active site will be one of these possibilities or one that has not yet been proposed is speculative; it is certain that the concept of an active site similar to that defined earlier will remain and will be a concept of fundamental importance in enzymology.

The chapters to follow in Part One are concerned with the general methods of enzyme mechanism study; Part Two is concerned with some particular examples that have been well worked out.

References

1. J. B. Sumner, *J. Biol. Chem.*, **69**, 435 (1926).
2. See J. H. Northrop, M. Kunitz, and R. M. Herriott, *Crystalline Enzymes*, 2nd ed., Columbia University Press, New York, 1948.
3. W. P. Jencks, *Catalysis in Chemistry and Enzymology*, McGraw-Hill, New York, 1969, p. 337.
4. For further reading on hydrogen bonds see G. C. Pimentel and A. L. McClellan, *The Hydrogen Bond*, Freeman, San Francisco, 1960.
5. For a more detailed discussion see G. Nemethy, *Angew. Chem. Int. Ed.*, **6**, 195 (1967).
6. For a more detailed exposition on the active site see D. E. Koshland, Jr., *Adv. Enzymol.*, **22**, 45 (1960).
7. H. Gutfreund, *Naturwissenschaften*, **54**, 402 (1967).

CHAPTER 2

Enzyme Methods

For those concerned with the study of enzyme mechanisms, the isolation, purification, and characterization of homogeneity of an enzyme are often looked upon as one distasteful, tedious task. Frequently, however, particularly in the study of similar enzymes from different sources, these mentioned tasks must be accomplished, and the care with which they are carried out might well determine the success or failure of the investigation. Therefore these tasks may not be neglected or treated lightly.

Because of their polypeptide nature and high molecular weights, enzymes cannot be treated as ordinary organic chemicals during characterization and purification. Indeed, they are often unstable in their own natural habitat—water—unless certain additives such as simple salts or particular metal ions are added.* Unfortunately, the manifestation of this sensitivity is not predictable and must be determined by trial and error. A good rule of thumb is to assume instability unless stability has been demonstrated. On the positive side of the ledger is the fact that a vast practical experience in enzyme purification has built up over the past 40 years, and this makes the selection of methods for isolation and purification somewhat simpler. The techniques that have been developed are generally applicable (with caution) to most enzymes since they are based on the common denominator of all enzymes, their protein constitution.

Isolation, purification, and characterization of purity and homogeneity will now be considered individually. These sections will be followed by a brief discussion of amino acid analysis and then by a short, exemplary enzyme purification and characterization of homogeneity. The techniques that are used overlap considerably, with the main difference being one of scale of the experiment. It should be kept in mind that most enzymes show a strong temperature dependence for denaturation and therefore, unless unavoidable, these processes should be carried out in the cold (0–5°C).

* Proteolytic enzymes frequently destroy themselves in a cannibalistic fashion (termed *autolysis*) because the enzyme protein may be its own substrate. However, proteolytic enzymes do not generally react with native proteins but save their talents for denatured proteins.

Isolation

Given a living species, the enzymes produced by the species are found in one of two locations—inside or outside the cell within which it was produced. Those inside are termed *intracellular* enzymes and those outside *extracellular* enzymes, and the extracellular enzymes are certainly the easier to isolate. For example, the pancreas in most mammals produces a juice that contains a variety of enzymes synthesized there and then excreted. These enzymes include proteases, lipases, and esterases. Many microorganisms produce enzymes which are excreted into the nutrient medium being fed them and are presumably useful in altering the nature of the nutrients so that the organisms might make better use of them. When this is the case, the isolation is very straightforward, as with extracellular mammalian enzymes. One simply collects the fluids involved, be it an excreted juice or a nutrient medium, by filtration or centrifugation of suspended solid contaminants. The procedure that follows will be discussed shortly.

Things are not so routine for intracellular enzymes. The source of enzyme production may be insect, fish, mammal, plant, or microorganism. The insect tissue, animal organ, or whole microorganism may be treated in one or more of the following ways: (1) extraction of the minced organ or insect part with water is often sufficient; (2) the tissue or microorganism suspension may be stirred at high speed (blender); (3) the suspension may be ultrasonicated before or after homogenization; (4) the suspension may be frozen and thawed; (5) the suspension may be treated with chemical "lysing" agents (i.e., reagents to rupture the cell wall); (6) the suspension may be treated with biological lysing agents, such as the enzyme lysozyme. All of these procedures have as their desired end the rupture of the cell membrane and the release of the intracellular components into the surrounding medium. From this point the treatment of these enzymes is similar to that of the extracellular enzymes, although they are often less stable.

An enzyme preparation isolated as discussed above (intracellularly or extracellularly) routinely contains many contaminants, such as nutrients, salts, nucleic acids, and carbohydrates. Thus a variety of methods have been worked out to remove the nonprotein constituents of the preparation. Among the methods used are fractional precipitation by organic solvents, by change of pH, and added salts. These methods take advantage of the properties that are a consequence of their fundamental chemical nature. Most high molecular weight polypeptides are inherently of limited solubility in high concentrations of organic solvents and inorganic salts. The solvents that have proven most useful are ethanol, acetone, and isopropanol, and the precipitations are carried out at very low temperatures ($0°$ to $-20°C$) to avoid solvent denaturation of the protein. The most commonly used salt is ammonium sulfate.

Fractionation by change of pH takes advantage of either the protein's limited solubility at or near its isoelectric point* or the limited solubility of certain contaminants at a given pH. For example, this method is occasionally useful in the precipitation of nucleic acid contaminants of a preparation since nucleic acids will readily precipitate at pH values near 5.5.

A very useful method of removing the bulk of nonprotein contaminants from an enzyme preparation is fractional adsorption of these contaminants on an insoluble support such as calcium phosphate gel or alumina C_γ gel.[1] Activated charcoal also is used occasionally as an adsorbent. This approach has found wide acceptance in enzyme isolation and purification because it frequently can be manipulated such that a high degree of fractionation of active enzyme from nonactive protein is achieved as well as the removal of nonprotein components of the broth. In general, the method involves adding a small, known amount of the gel to the preparation, centrifuging, separating supernatant and gel, and then adding more gel and repeating the process as often as necessary. At each stage both the gel fractions and the supernatant fractions are analyzed for the enzymic activity of interest.†

An example of an approach to the isolation and purification of the bacterial enzyme L-asparaginase from *Escherichia coli B*2 is given at the end of this chapter. In this case a high level of purification is essential because the enzyme is used in the treatment of human leukemias and hence it must be demonstrably free of any toxic contaminants. In mechanistic studies on enzymes the reliability of the data is obviously no better than the purity of the enzyme and one must be sure that there are no contaminants that interact with it.

Purification

Although numerous enzymes have been crystallized (over 100 and the number is growing), this is frequently a process that the enzyme chemist cannot utilize because he is working with a few milligrams or less of "pure" enzyme. Moreover, because of their large size, many protein and enzyme molecules tend to cocrystallize with impurities in the solution. This is further complicated by the fact that most proteins will tenaciously hold from 2 to 15% of their own weight in water of crystallization, even when dried under vacuum. Thus this exceedingly useful organic chemical method of purification is of limited use to the enzyme chemist for purification and useless as a measure of purity.

---

* See discussion of isoelectric points in the section on ion exchange later in this chapter.

† The analysis, or assay, depends on the enzyme. The choice of assay procedure is dictated by a combination of the considerations of its convenience and sensitivity. (See Chapter 4.)

The most common means of purification of an enzyme preparation after it has been successfully isolated involve one of two general types of column chromatography, *gel filtration* (gel permeation chromatography) and *ion exchange* chromatography. Both of these methods will be discussed in the following pages. In addition, the techniques of gel electrophoresis will be presented, followed by a brief discussion of the technique of isoelectric focusing. With the exception of ion exchange chromatography, these techniques have been developed largely within the past decade and have revolutionized analytical biochemistry.

a. Gel Filtration

Gel filtration is an exceedingly useful technique which provides a column chromatographic procedure for the fractionation of high molecular weight materials on the basis of their molecular size in solution. For a given type of biochemical (e.g., globular proteins), this is closely equivalent to a separation based on molecular weight.*

The process may be visualized with the aid of Figure 2-1. Suppose one

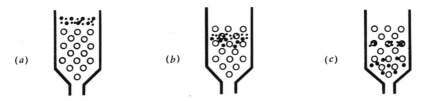

FIGURE 2-1. Schematic representation of a gel filtration experiment. (*a*) Beginning of the experiment—all particles moving at same rate. (*b*) Intermediate state—separation taking place due to preferential inclusion of smaller-sized particles in solution. (*c*) End of experiment separation is complete. (Figure courtesy of Pharmacia Fine Chemicals, Piscataway, N.J.)

has an enzyme preparation containing one important active component and contaminants that are of both larger and smaller molecular weights. After appropriate choice of gel material and column preparation, a sample of the enzyme preparation is placed on the column and the elution started. At this stage all particles in solution are traveling at the same rate (Figure 2-1*a*). However, the gel has been chosen so that the pore size of the porous gel particles allows the small contaminant to diffuse in entirely, the active component to diffuse in partly, and the large contaminant to diffuse in not

* In fact, accurate correlations of elution volume with molecular weight can be obtained for a homologous series of materials and this method has been used to assign approximate molecular weights to unknowns.

at all. Thus at an intermediate stage (Figure 2-1*b*) the small-sized contaminant is moving more slowly and the large-sized contaminant is moving more rapidly than the active component. At the end of the experiment (Figure 2-1*c*) a complete separation has been achieved and the materials elute from the column as discrete peaks which are collected in a fraction collector.*
In general, depending on the choice of gel and column conditions, a twofold difference in molecular weight is required for effective separation.

The gel materials used most frequently are the Sephadexes (Pharmacia Fine Chemicals, Piscataway, New Jersey) and the Bio-Gels (California Biochemical Corporation, Los Angeles, California). The Sephadexes are very high molecular weight dextrans which have been crosslinked and formed into a bead shape. The Bio-Gels are bead-shaped derivatives of polyacrylamide. Both types of material are white solids that absorb from one to about ten times their weight in H_2O to give a more or less dimensionally stable gel.

As illustrated in Figure 2-1, the mechanism of separation is based on a molecular sieving effect of the porous gel beads. This process has been theoretically quantitated through the following treatment.

Assume one has a column of total known volume V_t. This volume will consist of the following components when packed with a gel filtration medium: V_0, the void volume, or the volume of water outside the gel beads; V_i, the internal volume, or the volume of water inside the gel beads; and V_g, the volume of the gel matrix itself. Thus

$$V_t = V_0 + V_i + V_g \qquad (2\text{-}1)$$

Now V_0 is readily measured by eluting a substance known to be completely excluded by the gel particles being used, and V_i is calculated from the measured dry weight of the gel, a, and the water regain value of the particular gel particles according to Eq. 2-2. The W_r values are furnished by the gel supplier.

$$V_i = (a)(W_r) \qquad (2\text{-}2)$$

In the absence of molecular association of the substance being eluted with the gel beads the gel filtration process will be completely diffusion controlled. In this case the effluent volume (i.e., the volume required to elute the substance), V_e, will depend on the column parameters V_0 and V_i, and its distribution coefficient, K_d, between the gel and aqueous phases, according to Eq. 2-3.

$$V_e = V_0 + (K_d)(V_i) \qquad (2\text{-}3)$$

* This is an obvious oversimplification and most practical experiences are not this clean; the exception is desalting a protein.

Thus it is seen that K_d depends on column geometry only insofar as column geometry determines a particular V_0 and V_i. For example, K_d is independent of the height to diameter ratio.

Rearranging Eq. 2-3, we get Eq. 2-4 for K_d:

$$K_d = \frac{V_e - V_0}{V_i} = \frac{V_e - V_0}{(a)(W_r)} \tag{2-4}$$

Thus it is seen that K_d values may be calculated easily from a quantitative gel filtration experiment and will be a characteristic value for each substance on a particular type of gel. For molecules totally excluded from the gel, V_e will be equal to V_0 and K_d will be zero. For molecules completely included (assuming each diffusion process through the column goes to complete equilibrium) the V_e will equal the sum of V_0 and V_i, which may be seen from Eq. 2-1 to be equal to $V_i - V_g$. For such molecules, $K_d = 1$. If K_d is determined to be greater than 1, it means that V_i measured for the included compound is greater than that calculated from Eq. 2-2 and must be due to some type of adsorption complex being formed between the molecule and the gel material.

b. Ion Exchange

This is a very old process and has been used for years in commercial water demineralization. It did not find exceptional utility in biochemistry until adsorbents were developed that would not denature the sensitive biochemicals that were applied to them. Presently, there is a large variety of adsorbents that are useful for ion exchange chromatography of enzymes and other biochemicals, and among the most widely used are the substituted celluloses, such as carboxymethyl cellulose (CM-cellulose) and diethylaminoethyl cellulose (DEAE-cellulose). Figure 2-2a presents the general types of ion exchange materials available and Figure 2-2b shows the structure of the CM and DEAE groups.

Whereas in gel filtration the pH and ionic strength of the eluant are not critical (within normal ranges) these factors become crucial in ion exchange chromatography. The pH choice and control are critical because of the polyionic character of the protein. Consider the equilibrium in Eq. 2-5, where P is any normal protein:

$$H_3\overset{\oplus}{N}\text{-P-COOH} \underset{+H^{\oplus}}{\overset{-H^{\oplus}}{\rightleftharpoons}} H_3\overset{\oplus}{N}\text{-P-COO}^{\ominus} \underset{+H^{\oplus}}{\overset{-H^{\oplus}}{\rightleftharpoons}} H_2N\text{-P-COO}^{\ominus} \tag{2-5}$$

$$\text{I} \qquad\qquad\qquad \text{II} \qquad\qquad\qquad \text{III}$$

$$\text{Isolectric pH}$$

Resin Class	*Ionizing Group*
Strong acid resins	Sulfonate
	Resin-SO$_3^{\ominus}$ Na$^{\oplus}$; Resin-SO$_3$H
Weak acid resins	Carboxylate
	Resin-COO$^{\ominus}$ Na$^{\oplus}$; Resin-COOH
Strong base resins	Quaternized amine

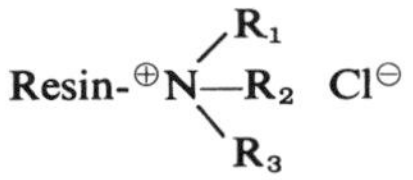

$$\text{Resin-}^{\oplus}\text{N}\underset{R_3}{\overset{R_1}{-}}R_2 \quad Cl^{\ominus}$$

Weak base resins Tertiary amine

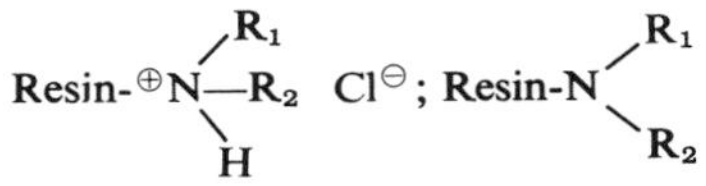

$$\text{Resin-}^{\oplus}\text{N}\underset{H}{\overset{R_1}{-}}R_2 \quad Cl^{\ominus}; \quad \text{Resin-N}\overset{R_1}{\underset{R_2}{}}$$

(*a*) General types available.

Sodium carboxymethylcellulose:

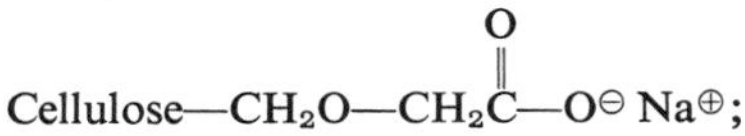

$$\text{Cellulose}-\text{CH}_2\text{O}-\text{CH}_2\overset{\overset{\text{O}}{\|}}{\text{C}}-\text{O}^{\ominus}\,\text{Na}^{\oplus};\ \text{CMC}$$

Diethylaminoethylcellulose:

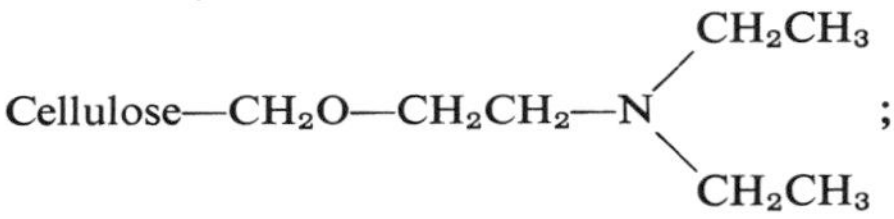

$$\text{Cellulose}-\text{CH}_2\text{O}-\text{CH}_2\text{CH}_2-\text{N}\overset{\text{CH}_2\text{CH}_3}{\underset{\text{CH}_2\text{CH}_3}{}}\ ;\ \text{DEAE-cellulose}$$

(*b*) Structures of two important groups in biochemical ion exchange.

FIGURE 2-2. Ion exchange materials.

At the precise pH value where **II** is the only species present, the molecule is electrically neutral and this is generally not a good pH for ion exchange chromatography. This pH is called the *isoelectric point* (pI) of the protein and is solely a function of the ionizable groups on the protein and their pK_a values. At pH values more acidic than pI the protein is positively charged (**I**) and therefore a cation exchange resin such as CM-cellulose is the adsorbent of choice. The opposite considerations hold at pH values more basic than pI (**III**) and here one uses an anion exchange resin. Since the resin functional groups of use in biochemistry are weak acids or bases, the choice of pH usually is not a trivial matter. Moreover, in solutions of low ionic strength, a protein's isoelectric pH is often its pH of minimum solubility.

The choice of ionic strength is also important, and the usual procedure is to start at a rather low level (generally in buffer at 0.005 to 0.01 M in buffer salts) and to gradually increase the ionic strength as the elution proceeds, a process called *gradient elution*, ordinarily reaching a final value of salt molarity of from 0.1 to 0.25 M.

The basic mechanism of ion exchange is illustrated in Eq. 2-6 and 2-7:

$$\text{resin}^{\ominus} \text{Na}^{\oplus} + \text{protein}^{\oplus} \rightleftharpoons \text{resin}^{\ominus\oplus} \text{protein} + \text{Na}^{\oplus} \tag{2-6}$$
$$\text{cation exchange pH} < \text{pI}$$

$$\text{resin}^{\oplus} \text{Cl}^{\ominus} + \text{protein}^{\ominus} \rightleftharpoons \text{resin}^{\oplus\ominus} \text{protein} + \text{Cl}^{\ominus} \tag{2-7}$$
$$\text{anion exchange pH} > \text{pI}$$

Recently, ion exchange adsorbents based on Sephadex and Bio-Gel have become available which provide superior abilities to resolve components differently from most common cellulose derivatives. This is presumably based on the greater exchange capacity and not on a combination of gel filtration and ion exchange, since the forces involved in ion exchange are approximately an order of magnitude stronger than those involved in retention by gel filtration. Moreover, there is no correlation between molecular size and pI for a protein.

c. Gel Electrophoresis

Electrophoresis is a process whereby a solution of material of interest is placed on an inert support and an electric current is applied to the sample, setting up an electric field. Since, in general, a protein molecule will have a finite charge in solution (except at pI), migration of the molecules will occur toward the pole of opposite charge. This simple phenomenon is the basis of a variety of methods of analytical electrophoresis, including paper electrophoresis, cellulose acetate strip electrophoresis, moving boundary electrophoresis, and polyacrylamide gel electrophoresis. Although each of these techniques has merit, the last, polyacrylamide gel electrophoresis, has found particular acceptance as a preparative method in protein purification studies. It is, moreover, reputed to be the most sensitive of all the analytical techniques for protein characterization.

The basis of separation in zone electrophoresis* lies in the mathematical equation that describes the mobility (velocity) of a particle in an electric field on a supporting medium (Eq. 2-8) where it may be seen that the rate

---

* The diagram in Figure 2-3 is applicable to any type of zone electrophoresis experiment. Zone electrophoresis is any electrophoresis done on an insoluble support, whereas that performed in solution is termed "moving boundary electrophoresis."

of migration is dependent upon, among other things, the overall charge Z on the particle, moving faster with a larger formal charge.[2]

$$u = \frac{Ze}{6\pi\eta a} = \frac{1}{E}\frac{dx}{dt} \qquad (2\text{-}8)$$

where Z = charge on the particle
$\quad\quad e$ = charge of an electron
$\quad\quad \eta$ = viscosity of the solvent
$\quad\quad a$ = radius of the particle
$\quad\quad E$ = electric field strength in volts per centimeter
$\quad\quad x$ = distance from origin in centimeters

Figure 2-3 is a schematic diagram of a gel electrophoresis experiment. After proper selection of conditions, the sample is applied to the gel, whose

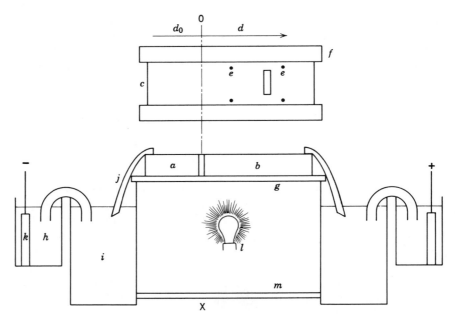

FIGURE 2-3. Block electrophoresis by moving-zone method. (*a*) Side view of starch block at time t_0; (*b*) initial zone; (*c*) top view of starch block after time t; (*d*) distance from origin to migrated zone; (*d*$_0$) distance from origin to substance with zero mobility (moved by solvent flow); (*e*) reference marker substances along each border; (*f*) strip used in molding block; (*g*) glass support; (*h*), (*i*) two-chamber electrode vessel; (*j*) connecting bridge to block; (*k*) electrode; (*l*) lamp; (*m*) connecting tube to equalize levels in electrode vessels. (After R. Trautman, in *Comprehensive Biochemistry* Vol. 7, M. Florkin and E. H. Stotz, eds., Elsevier, New York, 1963, p. 134.)

ends are connected to the current source through salt bridges between the buffer of choice and the electrode solution. The current is then turned on for a period of time and each component moves according to its charge. In a well-designed experiment all components will be well separated in the gel, which may then be sliced and the individual slices may be analyzed for the protein activity in question after elution.

d. Isoelectric Focusing

The apparatus for applying this technique has only recently become available and so its use and value have not yet been clearly demonstrated but, as will be seen, the potential is enormous, and this technique will surely take an important place in the biochemist's arsenal of macromolecule purification methods.

Isoelectric focusing may be simply defined as electrophoresis in a pH gradient. This means that if the electrophoresis is done within a pH range

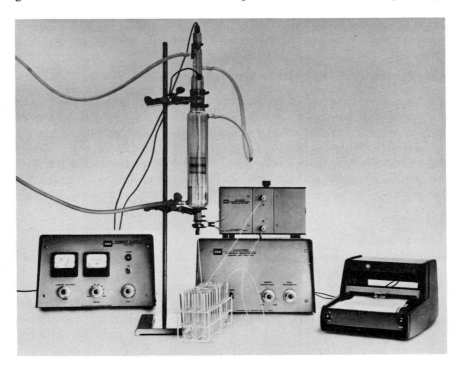

FIGURE 2-4. Isoelectric focusing equipment including power supply and specially designed column for electrofocusing experiments. Accessory fraction collector and uv monitor are also available. (LKB Instruments, Inc., Rockville, Md. 20852.)

that encompasses the pI of the component of interest, this component will move to the location in the gradient column where pH = pI and then stop, since at pI the component has no formal charge and therefore no tendency to migrate in an electric field. The gradient is formed by mixing together a series of zwitterionic buffer species of slightly (for a small pH span) or grossly different pK_a values, and then electrophoresing the mixture. Since the buffered species are zwitterionic, they will migrate from one another into the water in the column, but they will stop when sufficient ionization has occurred to give complete charge neutralization. This stopping point will differ for buffer species of different pK_a, and therefore a stable pH gradient may be established.

Figure 2-4 is an illustration of the apparatus, which is available in macro (440 ml) as well as micro (110 ml) scale.* One of the major advantages of this technique is that in addition to being a sensitive separation method, measurement of the pH in the region which contains the activity of interest reveals the material's isoelectric point (hence the name isoelectric focusing). Thus the method provides a value for pI in less than a day's time where other methods (such as moving boundary electrophoresis) often require weeks.

Characterization of Homogeneity and Purity

A most important part of the study of any enzyme is the determination of the extent of its purity. This is particularly true when one is working with an enzyme preparation that initially contained more than one type of activity or several activities of a similar type (e.g., proteinases and peptidases). In general, the purification steps are chosen so that enzyme activities of no interest are separated, but the certainty of this exclusion depends upon the sensitivity of the assay procedure used to monitor this unwanted activity.

a. Solubility

Once an enzyme preparation is thought to be pure, it may be subjected to a number of probes of its purity. As mentioned earlier, crystallization is not a satisfactory gauge of purity for a protein. On the other hand, solubility determination can be a good gauge if the experiments are carefully planned and executed. A pure protein will give a solubility curve as exemplified by the solid line of Figure 2-5. Depending on the extent of impurity of the sample, the solubility curve generated will deviate more or less from the solid line

---

* This equipment is made by LKB Instruments, Inc., Rockville, Md., 20852, who also make materials to be used in forming the gradient. Important references on this technique include Svensson, and Hagland.[3]

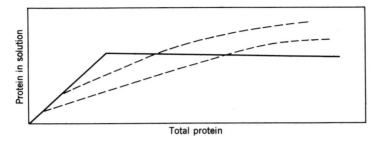

FIGURE 2-5. Solubility curves for pure and impure proteins. The solid line is observed for pure proteins, one of the dashed lines is observed for impure proteins.

shape and will appear more like one of the broken lines shown. Even with good technique a 2 or 3% impurity may be missed and, moreover, sizable amounts of protein are often required.

b. Electrophoresis

As mentioned at the start of this chapter, a macro scale technique for purification frequently is useful as a micro scale analytical procedure and this is certainly true of electrophoresis. A most penetrating probe of homogeneity is analytical gel electrophoresis on a polyacrylamide or starch gel at a series of pH values. Equipment now available allows this to be done in a short period of time, usually in less than one day. Besides extreme sensitivity in detecting contaminating species, this technique generally requires at most a few milligrams of enzyme and can often be performed on microgram quantities. The basis of the technique was discussed earlier in the section on purification.

c. Ultracentrifugation

This is a useful analytical technique, but it requires expensive equipment. With proper instrumentation it can be used on less than milligram amounts of protein.

The technique involves spinning the macromolecule sample in a very high speed centrifuge and observing the rate of migration of the macromolecule (in our case a protein) as a function of time at a constant rotational speed. The speeds attained can reach 60,000 rpm. By virtue of the optical means of detection provided on the instrument, a migrating component appears as a moving peak on a screen and can be photographed at various time intervals. A pure, homogeneous protein solution will give a series of photographs as

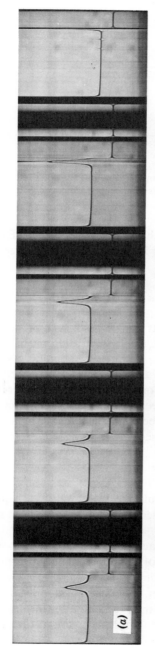

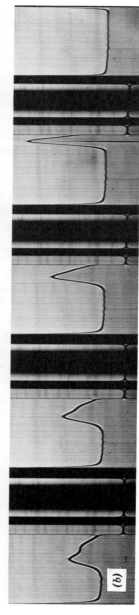

FIGURE 2-6. (a) Photographs of the UC pattern for a pure protein. Photographs were taken 8 minutes apart. (b) Photographs of the UC pattern for an impure (mixture) protein.

26

shown in Figure 2-6a, which shows a single symmetrical peak in each frame. A preparation with high molecular weight impurities may show a single unsymmetrical peak, a shoulder on an otherwise symmetrical peak, or multiple peaks, depending on the nature of the impurities. A mixture of proteins is shown in Figure 2-6b.

A series of photographs such as those in Figure 2-6a allows the calculation of a constant called the *sedimentation coefficient s*, which is defined in Eq. 2-9:

$$s = \frac{dx/dt}{\omega^2 x} \tag{2-9}$$

where x is the distance from the center of rotation to the center of the migrating peak, ω is the rotational velocity, and t is the time. The units of s are in cm/(sec)(dyne)(g) = sec, and now are reported in units of svedbergs (S) where $1\ S = 10^{-13}$ sec. The s is a characteristic constant for a given molecule under a given set of conditions and is usually reported as s_{20w}, which is the measured value of s mathematically corrected to the value that would be measured in water at 20°C.

The ultracentrifuge has also been extremely useful in the calculation of macromolecule molecular weights, which is done through the use of the Svedberg equation:[5,6]

$$M = \frac{RTs}{D(1 - \overline{V}\rho)} \tag{2-10}$$

where R is the gas constant, T is temperature (°K) of the experiment, D is the diffusion constant of the material under the experimental conditions (measurable in the ultracentrifuge), $\overline{V}$ is the macromolecule's partial specific volume (which for most proteins lies between 0.70 and 0.75 ml/g), and ρ is the density of the solution being measured.

Even though the techniques discussed above are widely used, none of them individually will firmly establish the homogeneity of a given sample. However, the manifestation of homogeneity in both ultracentrifugation and electrophoresis allows a fairly (but not completely) confident conclusion of purity, since one technique is based on the overall shape and size of the macromolecule and the other is based on its overall charge. Chromatography or electrophoresis on a number of supports (paper, cellulose acetate, etc.) in a variety of media (different pH, ionic strength, etc.) is also useful and often simple experimentally.

d. Amino Acid Analysis

Once a protein has been satisfactorily purified the determination of a reliable amino acid composition becomes an achievable and desirable goal. This information often allows for meaningful speculation about a protein's structure and provides a basis for comparison of the enzyme in question with other enzymes of similar catalytic properties whose amino acid composition has been determined. The elegant developmental work of Stein and Moore and their co-workers during the late 1940s and 1950s provided the basis for the sophisticated instruments available today.[7]

The technique of amino acid analysis generally involves the following steps. First, the sample is hydrolyzed to its constituent amino acids in any one of a variety of media, depending on what is desired in the analysis. Enzymic, alkaline, and acidic methods are all used, but by far the most common medium is $6N$ HCl, and the conditions of hydrolysis are 110°C for from 24 to 96 hours with the sample in a sealed evacuated tube. Second, the hydrolyzate is evaporated to dryness, the residue taken up in a small volume of buffer, and the solution placed on an ion exchange column. The adsorbent and elution conditions used have been worked out such that all the amino acids can be separated cleanly. Third, the eluate is analyzed in stream through the use of the ninhydrin reagent and the colored solution

$$\text{(2-11)}$$

$$\text{(2-12)}$$

$\lambda_{max} = 570$ nm

formed is quantitatively measured for absorbance at 570 nm. This reading allows for a quantitative determination of the amino acid composition of a protein. The reaction of ninhydrin with an amino acid is shown in Eqs. 2-11 and 2-12. The final product is purple in color.

It is difficult to find any reaction that has been more widely applied in analytical protein biochemistry.

The Purification of L-Asparaginase

Table 2-1 presents a simplified purification scheme for the important cancer chemotherapeutic agent L-asparaginase.[8] It is exemplary in that a variety of techniques have been applied. The authors started with a commercial sample of lyophilized (freeze-dried) cells of the microorganism *Escherichia coli B* and subjected them to sonication, to arrive at the first stage in the table, the crude extract. The resultant active material was then subjected to gel filtration (P-150 Bio-Gel), ion exchange chromatography (2 types, DEAE-cellulose and calcium hydroxylapatite) and, finally, preparative scale polyacrylamide gel electrophoresis. The fractionation patterns obtained in these steps are shown in Figures 2-7 to 2-10. Table 2-2 shows the amino acid composition. From ultracentrifugal analysis the molecular weight of the active enzyme was calculated to be 133,000 ($\pm$ 5000).

TABLE 2-1   Purification of *E. coli B* Asparaginase[a]

Step	Sample Volume (ml)	Total Protein (mg)	Total Units	Specific Activity (units/mg)	Stepwise Recovery (%)
Crude extract	140	12–14	3,500	0.2–0.25	
MnCl, heat	90–100	1,800	2,400	1.3	65
P-150 Bio-Gel	8–10	1,200	2,200	1.8	90
DEAE-cellulose	40–50	500–800	1,600	20–32	70
Calcium hydroxyl-apatite	25	3–6	800	150–250	50
Prep-disk electro-phoresis	4	1.5–2	600	300–400	90

[a] **From 20 g of lyophilized cells. Experimental details are given in the text. Recoveries in ultrafiltration and dialysis were essentially quantitative and these steps are omitted in the table. Overall purification of approximately 2000-fold, with 15% recovery.**

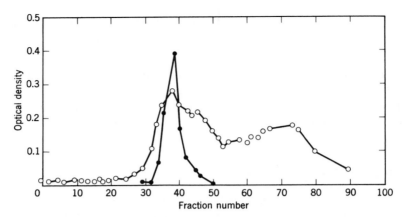

FIGURE 2-7. Gel filtration of asparaginase from heat-treated extracts of *E. coli B*. Bio-Gel P-150 was equilibrated with .02*M* potassium phosphate buffer (pH 8.0) and the same buffer was used for elution at 60 ml/hr. Protein concentration (open circles) is in terms of optical density at 280 nm; asparaginase activity (closed circles) in terms of optical density at 500 nm.

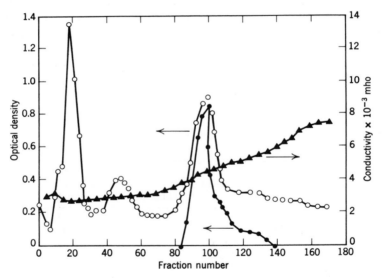

FIGURE 2-8. DEAE-cellulose chromatography of asparaginase fraction from gel filtration. A 3.0 × 30 cm column was equilibrated with .02*M* potassium phosphate buffer (pH 8.0). After about 1.5 column volumes of starting buffer, a linear gradient to 0.2*M* potassium phosphate buffer (pH 6.35) was started at the point indicated by an arrow. The column was operated at approximately 80 ml/hr, collecting 10-ml fractions. Protein concentration (open circles) in terms of optical density at 280 nm, asparaginase activity (closed circles) in terms of optical density at 500 nm, and conductivity is represented by solid triangles.

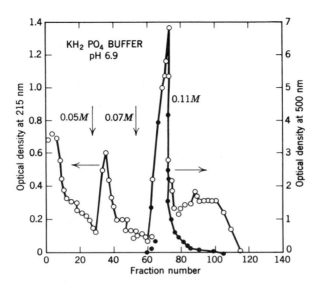

FIGURE 2-9. Calcium hydroxylapatite chromatography of *E. coli* asparaginase fraction from DEAE-cellulose column. A 2 × 20 cm column was equilibrated with $0.05M$ potassium phosphate buffer (pH 6.9). The buffer was changed to 0.07 and $0.11M$ potassium phosphate buffers (pH 6.9), respectively, at the points indicated by arrows. Protein concentration (open circles) is in terms of optical density at 215 nm, and asparaginase activity (closed circles) in terms of optical density at 500 nm. The column was operated at about 50 ml/hr, and 5-ml fractions were collected.

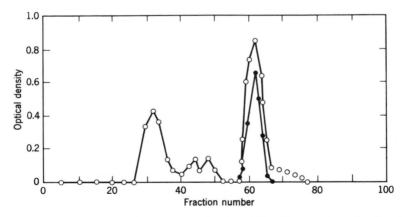

FIGURE 2-10. Polyacrylamide gel electrophoresis of *E. coli* asparaginase fraction from calcium hydroxylapatite column. (See text for preparation of column.) The column is operated for 10 hr at 4 mA at 0° and eluted with $0.05M$ Tris buffer (pH 8.8) collecting 5-ml fractions. Protein concentration (open circles) is in terms of optical density at 215 nm and asparaginase activity (closed circles) is in terms of optical density at 500 nm.

TABLE 2-2   Amino Acid Composition of *E. coli B* Asparaginase

Amino Acid	Amino Acid/His		Approximate Minimum Number of Residues	Residues 133,00 mol. wt.
	24 hr	72 hr		
Asp	15.1	14.0	15	180
Thr	9.3	8.2	10	120
Ser	4.5	4.4	5	60
Glu	6.8	6.6	7	84
Pro	3.7	3.6	4	48
Gly	9.1	8.9	9	108
Ala	9.7	9.3	10	120
Val	9.4	10.2	10	120
Cys[a]			(0.5)[a]	6
Met	1.9		2	24
Ile	3.6	3.6	4	48
Leu	7.2	7.0	7	84
Tyr	3.3	3.1	4	48
Phe	2.6	2.5	3	36
Lys	6.7	6.1	7	84
His	(1)	(1)	(1)	12
Arg	2.7	2.4	3	36
Trp[b]			1	12

[a] Cys determined by the method of Ellman, *Arch. Biochem. Biophys.*, **82**, 70 (1959), gave approximately 6 half-cystines/133,000 mol. wt.

[b] Trp determined by the method of Beaven and Holiday, *Adv. Protein Chem.*, **7**, 375 (1952). Minimal mol. wt., 22,170.

References

1. See M. Dixon and E. C. Webb, *Enzymes*, 2nd ed., Academic Press, New York, 1964, p. 42, for a preparation of these gels and a further discussion of this subject.
2. R. Audubert and S. deMende, *The Principles of Electrophoresis*, Macmillan, New York, 1959, p. 68. As presented, this expression assumes no interaction between ionic species in solution. In theory, such interactions usually decrease a particle's mobility.
3. H. Svensson, *Acta Chem. Scand.*, **15**, 325 (1961); **16**, 456 (1963); H. Hagland, *Science Tools*, **14** (2), 1 (1967).
4. H. K. Schachman, *Ultracentrifugation in Biochemistry*, Academic Press, New York, 1959.
5. T. Svedberg, *Kolloid-Z.*, **36**, 53 (1925).
6. H. K. Schachman and M. A. Lauffer, *J. Am. Chem. Soc.*, **72**, 4266 (1950).
7. D. H. Spackman, W. H. Stein, and S. Moore, *Anal. Chem.*, **30**, 1190 (1958), and references cited therein.
8. H. A. Whelan and J. C. Wriston, Jr., *Biochemistry*, **8**, 2386 (1969).

CHAPTER 3

Chemical Modification Studies

In Chapter 1, a definition was given for the active site of an enzyme molecule, and this concept of a crucial region of the molecule where reactive side chains are positioned with a delicate interdependence is of fundamental importance in enzymology. The reactive catalytic groups are generally side chains from amino acid residues of the peptide backbone, and their identity is an obvious requirement for even an elementary understanding of the enzymic process. The most widely used approach to the identification of these residues is the use of one or more of the chemical modification methods to be discussed in this chapter.

It is easy to see the value of modification experiments if one first considers the simple case of the reaction of one molecule of reagent with one side chain group on the enzyme molecule. First, let it be said that this one-to-one relationship is by no means to be expected, since in general there will be in the protein more than one residue of the particular amino acid whose side chain is reacting. Therefore, if the reaction occurs in a one-to-one fashion, the researcher is fortunate. We are even more fortunate if such a reaction leads to the complete loss of enzymic activity, since this clearly identifies the particular amino acid modified as a crucial one in the catalytic process. A partial loss of activity would mean that the side chain that had been modified was near the active site and possibly involved in noncovalent binding interactions but was not necessary for catalysis to occur. Its modification, then, would have disrupted, but not drastically, the integrity of the active site. It is conceivable, of course, that a modification of a noncatalytic group could so seriously affect the solution geometry of the active site as to make the enzyme nonfunctional. Therefore one should try more than one approach to the modification of a particular residue and, moreover, attempts should be made through measurements such as sedimentation constant, electrophoretic mobility, and gel filtration behavior to demonstrate the basic retention of three-dimensional structure after modification.

No loss of activity in the one-to-one example given would mean that the modified residue was not contributing in any observable way to the enzymic catalysis.

This procedure rapidly becomes more complex when more than one residue of a particular amino acid in an enzyme is modified. Not only does one have to worry about which (if any) of the modified residues are catalytically important, but one must also take into account the probable disruption of the enzyme's three-dimensional structure in solution caused by the more extensive modification. One technique that has been developed to aid this situation is the use of a "competitive inhibitor" (see Chapter 6) in a first modification step followed by a repeat modification step after removal of the inhibitor. (This second step is often carried out using radioactive reagent.) The competitive inhibitor blocks susceptible residues at the active site. Therefore a correlation of enzymic activity before and after each modification and a determination of the number of residues modified in each step helps one determine the effect on the active site and the number of residues of the susceptible type in the active site region.

This chapter will first treat the general types of modification reactions that are commonly used in present-day enzymology[1,2] and then will present some specific examples where judicious choice of reagent has led to the one-to-one reaction discussed previously. This is termed *covalent labeling* of the active site.[3-6] The chapter will conclude with a brief description of the use of "reporter groups" in chemical modification studies, which is a relatively new technique of considerable potential.

Before proceeding, however, a few precautions must be noted. Although the reactions to be discussed normally proceed by the expected organic reaction mechanism (S_N2, Michael addition, etc.), the observed reactivity patterns of organic chemistry are generally not useful because of two complicating factors. First, the binding (specificity) restrictions imposed by the enzyme upon its substrates may not allow close enough approach of the modifying reagent and the residue to occur. Second, the electronic and steric conditions present at the location on the protein being modified (the reacting site) may be quite unlike the simple organic analog to the reaction in question. Thus the nature of the microenvironment in which the reaction is to occur may accelerate it, inhibit it, or have no effect when compared to the same reaction run in a normal organic procedure.*

These two considerations and the fact that the native protein structure may or may not be sensitive to the sometimes extreme (for proteins) conditions used mean that unless the proper controls are run on a given modification experiment, the results may easily be misleading and misinterpreted. Among

---

* For example, there are 13 tyrosine residues in the bacterial enzyme subtilisin Carlsberg. Upon treatment with I_2, 3 to 4 are monoiodinated, 2 to 3 are diiodinated, and the remainder are unreactive. Free tyrosine in solution under the same conditions would be completely diiodinated.[7]

the points that must be considered in any modification study are (1) the occurrence of peptide bond cleavage, (2) changes in all types of enzymic activity known to be present, (3) reaction condition variation (i.e., changes in concentration ratios, rate of addition, order of addition, etc.), (4) effect of buffer species on the modification and on the modified protein, and (5) proper amino acid analysis of the modified protein.

General Modification Reactions

a. Acylating Reagents

Acylation reactions on proteins are commonly applied because they are rapid and are usually simple to carry out. They occur at room temperature over a broad range of pH (4–9). The acylating agents are usually highly reactive; therefore the reactions are generally nonspecific and occur at a number of sites on the protein molecule. The functional groups on the protein that are reactive toward acylating agents are aliphatic and aromatic hydroxyl groups, amino groups, and sulfhydryl groups. Among the reagents commonly used are acetic and succinic anhydride, acetylimidazole, ethyl thioltrifluoro-acetate, O-methyl *iso*urea, imidate esters, and sulfonyl halides. A generalized reaction scheme is presented in Eq. 3-1, with **I** being the generalized product.

$$\text{R—C—X + protein}\begin{cases}\text{—OH}\\\text{—SH}\\\text{—NH}_2\\\text{—} \bigcirc \text{—OH}\end{cases} \longrightarrow \text{protein}\begin{cases}\text{—O—C—R}\\\text{—S—C—R}\\\text{—NH—C—R}\\\text{—}\bigcirc\text{—OC—R}\end{cases} + \text{HX}$$

I   (3-1)

If the protein functional group can ionize, for example, $RNH_3{}^+$—RNH_2, the dissociated form is usually much more reactive. This explains why the guanidino group of arginine cannot be acylated; it is so strongly basic ($pK_a \geq 12$) that under the usual acylation conditions it is completely protonated.

The reagent O-methyl *iso*urea and the imidate esters are worthy of mention since, in contrast with the other acylating agents, these reagents (which react primarily with amines) lead to retention of the charge type. Other acylating agents generally form amides with amines, but these form guanidino and amidino groups, respectively. The reactions are shown in Eqs. 3-2 and 3-3. The imidate esters are prepared from nitriles plus ethanolic HCl.

$$Pr—NH_2 + HN{=}\overset{\overset{\displaystyle OCH_3}{|}}{C}—NH_2 \longrightarrow Pr—NH—\overset{\overset{\displaystyle NH}{\|}}{C}—NH_2 + MeOH \qquad (3\text{-}2)$$

$$Pr—NH_2 + R—\overset{\overset{\displaystyle NH}{\|}}{C}—OEt \longrightarrow Pr—NH—\overset{\overset{\displaystyle NH}{\|}}{C}—R + EtOH \qquad (3\text{-}3)$$

Perhaps the most important "acylating" agents have been diisopropyl-fluorophosphate (**II**) and phenylmethanesulfonyl fluoride (**III**). These reagents are actually phosphorylating and sulfonylating agents, respectively. They

have proven to be effective covalent labels for the active sites of the class of enzymes called *serine* hydrolases and the reaction of these materials with these serine enzymes will be considered later in this chapter.

The mechanisms of all these acylations involve nucleophilic attack by groups on the protein upon the reactive carbonyl, sulfonyl, or phosphoryl center of the acylating reagent, followed by rapid breakdown of the inter-mediate formed. These intermediates are presumably pentacovalent for reagents **II** and **III** and tetrahedral for carbonyl-type reagents.

b. Arylating and Alkylating Reagents

These reagents are compounds with an active halogen atom that is suscept-ible to nucleophilic displacement and react via an S_N2 mechanism. By far the most common arylating reagent is 2,4-dinitrofluorobenzene (**IV**), which not only has been used as a modifying reagent but has found wide use in the determination of protein N-terminal amino acids. It reacts with α- and ε-amino groups in proteins to form the N-dinitrophenyl derivatives. Structure **IV** is important historically, since it was extensively used by Sanger in his Nobel Prize-winning studies on the structure of insulin.[8]

The most common alkylating agents have been α-halo acids, amides, and ketones such as iodoacetate, chloroacetamide, and phenacyl halides. The α-halo acids and amides have been found to react most frequently with sulf-hydryl groups since the conditions are usually arranged to preclude reaction with amino groups. This is shown in Eq. 3-4 with iodoacetic acid as the reagent.

$$I—CH_2COO^- + Pr—SH \longrightarrow Pr—S—CH_2COO^- + HI \qquad (3\text{-}4)$$

Phenacyl halides have usually been found to react with methionine groups, according to Eq. 3-5, to give sulfonium salts.*

$$Pr—CH_2—S—CH_3 + \phi\overset{\overset{\displaystyle O}{\|}}{C}CH_2Br \longrightarrow Pr—CH_2—\underset{\underset{\displaystyle Br^{\ominus}}{\oplus}}{\overset{\overset{\displaystyle CH_2—\overset{\overset{\displaystyle O}{\|}}{C}—\phi}{|}}{S}}—CH_3 \qquad (3\text{-}5)$$

Phenacyl halides have also been found to react with a carboxyl group in pepsin, with the formation of the appropriate ester (Eq. 3-6). This is not surprising since phenacyl halides are derivatizing reagents for carboxylic acids.

$$\overset{\overset{\displaystyle O}{\|}}{\phi C}—CH_2Br + Pr—COO^- \longrightarrow \overset{\overset{\displaystyle O}{\|}}{\phi C}—CH_2—O—\overset{\overset{\displaystyle O}{\|}}{C}—Pr + Br^- \qquad (3\text{-}6)$$

α-Haloketones have found particular use in the derivatization of histidine side chains in a one-to-one fashion and will be discussed in greater detail later in this chapter.

c. Reagents with Activated Double Bonds

The two most widely used reagents in this category are N-ethylmaleimide (**V**) and acrylonitrile (**VI**).[1] These reagents have both proven quite useful for the modification of sulfhydryl groups, as shown in Eqs. 3-7 and 3-8.

---

* One case has been reported where the evidence favors the formation of a sulfonium ylide rather than a sulfonium salt from the reaction of a phenacyl bromide with a chymotrypsin methionine.[9]

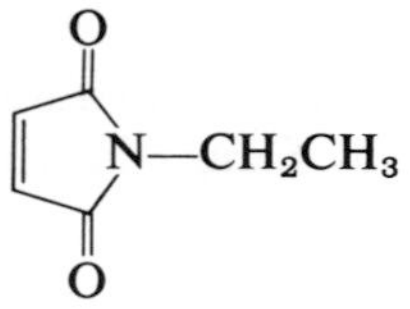

V   VI

$$\text{Pr—SH} + \quad \xrightarrow{\quad} \quad (3\text{-}7)$$

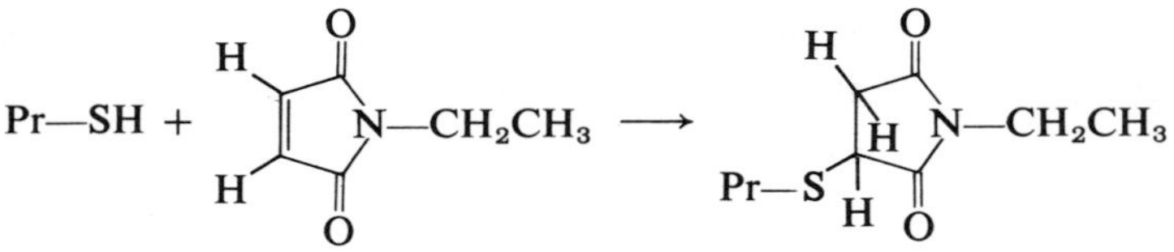

$$\text{Pr—SH} + \text{CH}_2\!\!=\!\!\text{CH—C}\!\equiv\!\text{N} \longrightarrow \text{Pr—S—CH}_2\text{CH}_2\text{—C}\!\equiv\!\text{N} \qquad (3\text{-}8)$$

d. Electrophilic Reagents

Species that are electron deficient have not been widely used in protein modification studies due to their tendency to oxidize certain susceptible groups on the protein. For example, one widely used reagent for electrophilic substitution in proteins is I_2, which most commonly reacts with tyrosine to form mono- or diiodotyrosine via a normal electrophilic aromatic substitution mechanism. This reagent is also known to be able to oxidize both cysteine and tryptophan side chains and, moreover, to iodinate histidine side chains in certain cases, and therefore the utmost care must be taken when using it.

A much more useful electrophilic reagent has been developed by Vallee and co-workers:[10,11] tetranitromethane (**VII**). This reagent has been found to be essentially specific for tyrosine residues in proteins, leading to the

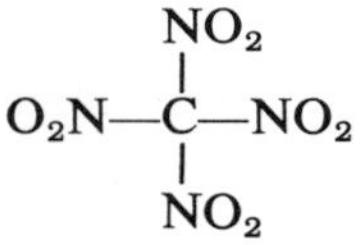

VII 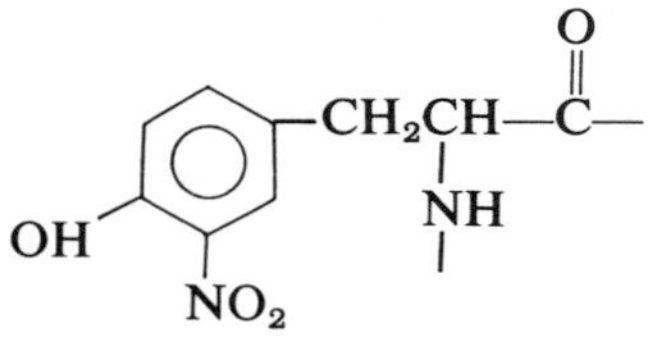   VIII

formation primarily of *mono*nitrotyrosine (**VIII**). Since **VIII** has a distinct yellow color above pH 7.2, the reaction is very useful and relatively easy to quantitate spectrophotometrically. The nitration of tyrosine in proteins is said to be quantitative.[10,11] Studies by Bruice and co-workers[12,13] on simple phenols and alcohols suggest that this may not always be the case; therefore care must be taken when using this reagent to accurately determine the extent of nitration and the identity of the nitrated residues.

e. Oxidizing Reagents

Although hydrogen peroxide and N-bromosuccinimide have found some acceptance as oxidizing agents in protein chemistry, the variability of susceptibility of functional groups as a function of the protein under study, combined with the ever-present danger of oxidative peptide bond cleavage, have made chemical oxidation studies very difficult to carry out. On the other hand, the sensitized photooxidation of proteins has been carried out routinely in a number of systems. In general, one uses methylene blue dye or rose bengal dye as the sensitizer and a visible light source as the energy supply. Although methionine, cysteine, tryptophan, histidine, and tyrosine may be oxidized (depending on conditions), remarkable specificity has often been observed in a photooxidation experiment. In some cases the technique has allowed the stoichiometric oxidation of one mole of histidine or cysteine per mole of enzyme and accompanying loss of enzymic activity.[14]

f. Reducing Reagents

The disulfide linkage of cystine is the only group in protein molecules readily susceptible to reduction, and it is not present in all proteins. The most widely used reducing agent at present is 2-mercaptoethanol, which is generally used in large excess over disulfide groups to ensure maintenance of the reduced form of the product. The reaction is shown in Eq. 3-9.

$$\text{Pr}\text{—}\text{S}\text{—}\text{S}\text{—}\text{Pr} + \text{H}\text{—}\text{S}\text{—}\text{CH}_2\text{CH}_2\text{OH} \longrightarrow \underset{\underset{\text{CH}_2\text{CH}_2\text{OH}}{|}}{\overset{}{\underset{|}{\text{Pr}\text{—}\text{S}}}} + \text{HS}\text{—}\text{Pr} \qquad (3\text{-}9)$$

HSCH₂CH₂OH

$$\text{Pr}\text{—}\text{SH} + \underset{\underset{\text{OH}}{|}}{\text{HOCH}_2\text{CH}_2\text{—}\text{S}\text{—}\text{S}\text{—}\text{CH}_2\text{CH}_2}$$

The use of mercaptoethanol is gradually being supplanted by the use of two anomeric dithiols developed by Cleland,[15] dithiothreitol and dithioerythritol **(IX)**. The reaction is shown in Eq. 3-10. As may be seen in Eq. 3-10b, the

$$\text{HS}\text{—}\text{CH}_2\text{—}\underset{\underset{\text{OH}}{|}}{\text{CH}}\text{—}\underset{\underset{\text{OH}}{|}}{\text{CH}}\text{—}\text{CH}_2\text{SH}$$

IX

intramolecular nature of the second step not only removes the requirement for a large excess of reducing agent but also makes this step very rapid and prevents reversal of Eq. 3-10a.

$$Pr-S-S-Pr + IX \longrightarrow Pr-SH + Pr-S-S-CH_2CH-CH-CH_2SH$$
$$\underset{\displaystyle OH \quad OH}{|\qquad |}$$

$$(3\text{-}10a)$$

$$(3\text{-}10b)$$

The examples discussed thus far illustrate the variety of reagents and reaction types that have been useful in chemical modification studies on proteins and enzymes. These examples, however, are by no means complete with respect to reagents for a given type of modification. Table 3-1 provides a summation of the specificity characteristics of, and references to, a large number of useful reagents for enzyme and protein modification. This compilation shows that few of these reagents are specific for only one type of residue and recalls, therefore, the precautions that must be observed when a chemical modification is performed.

Covalent Labeling of the Active Site

In this book a reagent will be considered to be a covalent label for an enzyme's active site if it can be shown that not more than two moles of reagent have been introduced per mole of enzyme active sites with concomitant substantial loss of enzymic activity. In the ideal case, which fortunately has been observed for a number of enzymes, one equivalent of reagent reacts per active site of enzyme and leads to 100% inactivation of the enzyme. When this occurs one is in a very good position to infer something about the catalytic mechanism of the enzyme, since, assuming proper controls are run, the stoichiometric inactivation of an enzyme by one mole of a covalent label clearly implicates the residue labeled as being involved in the chemical process of catalysis.

In general, one approaches this task of covalently labeling the active site of an enzyme only after sufficient data are known about the enzyme's specificity and its elementary kinetic behavior. The specificity studies determine the molecular "preferences" of the enzyme with respect to size and shape of the substrate and the type of reaction catalyzed. The kinetic and specificity studies both give clues about the actual nature of the catalytic groups involved.

With this information and a fair measure of chemical intuition a molecule may then be designed which will conform to most or all of the enzyme's known specificity requirements and will have the additional feature of having an unusual reactive group. This molecule, when allowed to interact with the enzyme, will presumably be directed to the active site of the molecule by virtue of its similarity to common substrates. The function of the reactive group of the labeling molecule is to react with one of the catalytic functionalities of the enzyme as the enzyme attempts its normal catalysis. Because of the unusual nature of this reactive group, however, the catalytic process stops once the initial reaction occurs, as the intermediate formed is stable to the rest of the catalytic process. A number of examples will amply illustrate this approach.

Consider the enzyme pepsin. Specificity studies indicated as early as 1951 that the most rapidly hydrolyzed substrates were of the general structure **X**.[16] Moreover, kinetic studies had shown that the pH optimum for these substrates was approximately 2.0, leading to the suggestion that carboxyl groups were the catalytically active species. Thus reagents **XI** and **XII**, *N*-diazoacetyl-norleucine methyl ester and L-1-diazo-4-phenyl-3-tosylamidobutanone-2, respectively, were designed as potential covalent labels. The diazomethyl-carbonyl moiety (underlined) had long been known to be highly reactive to carboxyl groups in the presence of Cu(II), and these compounds (particularly

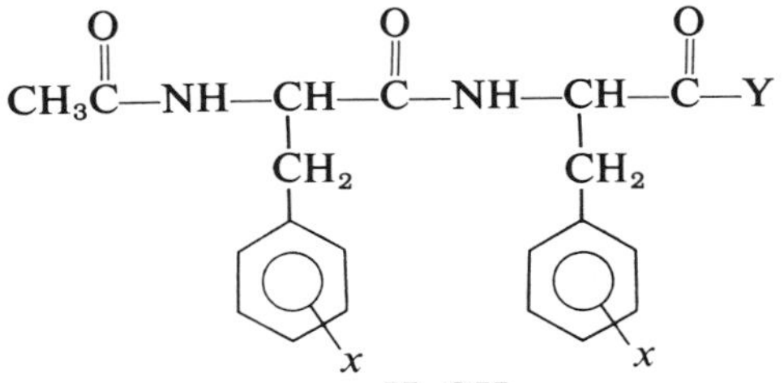

$x = $ H, OH
$Y = $ OH, OR, NH_2

X

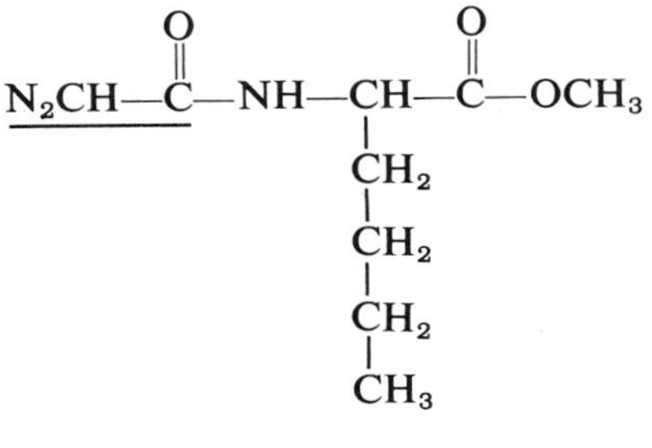

XI

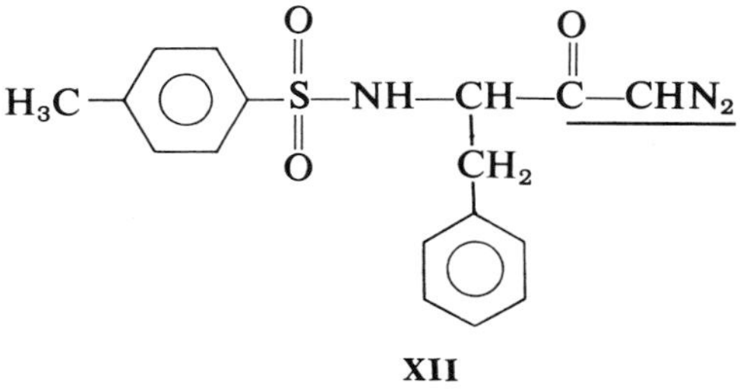

XII

TABLE 3-1 Specificity of Reagents Commonly Used to Chemically Modify Proteins[a]

	Peptide Gln, Asn	α-Amino, Lys	α-Carboxyl Glu, Asp	Cystine	Arg	Ser, Thr	His	Trp	Tyr	Cysteine	Met	Reference[b]
Acetic anhydride		+				+			+	+		1, 2
Acetyl imidazole		±							+	+		3
Acrylonitrile		+								+		4
Acyl anhydrides		+				±			+	+		1, 2
Acyl chlorides		+				+			+	+		1, 2
Aldehydes, formaldehyde	+	+			+		+	±		+		1, 2
Aryl sulfonyl chlorides		+				±	+		+	+		1, 2
Azide		+								+		5
Azlactones		+								+		6
Bromethyl amine										+		7
Bromopicrin, chloropicrin										+		8
Bromopyruvate										+		9
N-Bromosuccinimide							+	+	+	+		10
Carbodiimides			+									11
Carbon disulfide		+										12
Carbon suboxide		+							+	+		1, 2
N-carboxyanhydrides		+										13
Chlorotrifluoroquinone		+										14
Cyanate		+				+				+		15
Cyanuric fluoride									+			16
Diacetyl		+			+							17
N,S-Diacetyl-2-aminoethane-thiol		+										18
Dialdehydes, diketones		+			+							19
Dialkylfluorophosphates		+				±		±				20
Diazomethane, diazoacetimide		+	+						+			1, 2
Diazonium salts		+			+		+	+	+	+		21
Diethylpyrocarbonate		+					+		+	+		22, 23
Diketene		+										24

Reagent									Ref.
Dimethyl sulfate	+				+	+			1, 2
Diphenyldiazomethane					+	+			25
Dithionite				+					26
Ethanethiol trifluoroacetate	+								27
Ethyl acetimidate, imido esters	+							±	28
N-Ethylmaleimide	±							+	1, 2, 29
Ferricyanide, iodosobenzoate				+			+	+	1, 2
Fluorodinitrobenzene	+				+	+		+	30
Guanyldimethylpyrazole nitrate	+								31
Haloacetate, amide	+				+		+	+	+	1, 2
Heavy metals	±				±			±	+	1, 2
Hydrogen peroxide								±		32
Hydroxylamine					+		+			33
Hydroxynitrobenzyl bromide							+			34
Iodine-KI, iodine monochloride	+				±		+	+	+	1, 2, 35
Isocyanate	+		±				+	+		36
Ketene	+		+				+	+	+	37
Malonaldehyde	±					+				38
Mercuribenzoate	±				±	+		+	+	1, 2
Methanol-HCl, thionyl chloride	+			+					39
2-Methoxy-5-nitrotropone	+						+	+	40
Methyl iodide	+			+		+	+	+	1, 2
O-Methylisourea	+								1, 2
Mustards	+					+		+	+	1, 2
Naphthoquinone sulfonate	+								41
Ninhydrin	+				±	+	+	+	42
Nitrophenyl acetate	+			±		+		+	43
Nitrous acid	+			+			+	+	1, 2
Oxazolones	+			+				+	1

(continued)

43

TABLE 3-1 (*continued*)

	Peptide Gln, Asn	α-Amino, Lys	α-Carboxyl, Glu, Asp	Cystine	Arg	Ser, Thr	His	Trp	Tyr	Cysteine	Met	Reference[b]
Ozone								+				44
Performic acid				+			±	+		+	+	45
Peroxide-dioxane								+		+	+	46
Peroxyacetyl nitrate									+	+		47
Phenylisothiocyanate		+							±	+		48
Phosphorus oxychloride		+		+					+	+	±	1, 2
Photooxidation							+	+	+	+		49
Propylene, ethylene oxide		+	+				+	+	+	+	±	1, 2, 50
Tetranitromethane									+	+	±	51
Thioglycolate, cyanide, sulfite				+								1, 2
Trinitrobenzene sulfonate		+										52

[a] Source: **B. L. Vallee and J. F. Riordan, *Ann. Rev. Biochem.*, 38, 733 (1969).**
[b] References as follows:

1. R. M. Herriott, *Adv. Protein Chem.*, **3**, 169 (1947).
2. F. W. Putnam, *The Proteins*, 1st ed., Vol. 1, Part B, Academic Press, New York (1953) p. 893.
3. J. F. Riordan, W. E. C. Wacker, and B. L. Vallee, *Biochemistry*, **4**, 1758 (1965).
4. J. P. Riehm and H. A. Scheraga, *Biochemistry*, **5**, 93 (1966).
5. H. S. Olcott and H. Fraenkel-Conrat, *Chem. Rev.*, **41**, 151 (1947).
6. J. DeJersey, M. T. C. Runnegar, and B. Zerner, *Biochem. Biophys. Res. Commun.*, **25**, 383 (1966).
7. H. Lindley, *Nature*, **178**, 647 (1956).
8. E. Fredericq and V. Desreux, *Bull. Soc. Chim. Biol.*, **29**, 100, 105 (1947).
9. H. P. Meloche, *Biochemistry*, **6**, 2273 (1967).
10. B. Witkop, *Adv. Protein Chem.*, **16**, 221 (1961).
11. D. G. Hoare and D. E. Koshland, Jr., *J. Biol. Chem.*, **242**, 2435 (1967).
12. A. L. Levy, *J. Chem. Soc.*, **1950**, 404.
13. M. Sela and E. Katchalski, *Adv. Protein Chem.*, **14**, 392 (1959).
14. K. Nakaya, H. Horinishi, and K. Shibata, *J. Biochem. (Tokyo)* **61**, 337 (1967).
15. G. R. Stark, *Biochemistry*, **4**, 588, 1030 (1965).
16. K. Kurihara, H. Horinishi, and K. Shibata, *Biochim. Biophys. Acta*, **74**, 678 (1963).

17. J. A. Yankeelov, Jr., C. D. Mitchell and T. H. Crawford, *J. Am. Chem. Soc.*, **90**, 1664 (1968).
18. J. Baddiley, R. A. Kekwick, and E. M. Thain, *Nature*, **170**, 968 (1952).
19. T. P. King, *Biochemistry*, **5**, 3454 (1966); K. Nakaya, H. Horinishi, K. Shibata, *J. Biochem. (Tokyo)* **61**, 345 (1967).
20. E. R. Jansen, M. D. F. Nutting, R. Jang, and A. K. Balls, *J. Biol. Chem.*, **179**, 189 (1949).
21. K. Landsteiner, *The Specificity of Serological Reactions*, 2nd ed., Harvard University Press, Cambridge, Mass., 1945.
22. C. G. Rosen and I. Fedorcasak, *Biochem. Biophys. Acta*, **130**, 401 (1966).
23. A. Muhlrad, G. Hegyi, and G. Toth, *Acta Biochim. Biophys.*, **2**, 19 (1967).
24. A. Marzotto, P. Pajetta, and E. Scoffone, *Biochem. Biophys. Res. Commun.*, **26**, 517 (1967).
25. G. R. Delpierre and J. S. Fruton, *Proc. Nat. Acad. Sci. U.S.*, **56**, 1817 (1966).
26. W. Windus and H. G. Turley, *J. Am. Leather Chemists Assoc.*, **36**, 603 (1941).
27. R. F. Goldberger and C. B. Anfinsen, *Biochemistry*, **1**, 401 (1962).
28. M. J. Hunter and M. L. Ludwig, *J. Am. Chem. Soc.*, **84**, 3491 (1962).
29. C. F. Brewer and J. P. Riehm, *Anal. Biochem.*, **18**, 248 (1967).
30. F. Sanger, *Biochem. J.*, **39**, 507 (1945).
31. A. F. S. A. Habeeb, *Biochim. Biophys. Acta*, **34**, 294 (1959).
32. K. Freudenberg, W. Dirscherl, and H. Eyer, *Z. Physiol Chem.*, **202**, 128 (1931).
33. H. Fraenkel-Conrat, *Arch. Biochem.*, **28**, 452 (1950).
34. D. E. Koshland, Jr., Y. D. Karkhanis, and H. G. Latham, *J. Am. Chem. Soc.*, **86**, 1448 (1964).
35. M. E. Koshland, F. M. Englberger, M. J. Erwin, and S. M. Gaddone, *J. Biol. Chem.*, **238**, 1343 (1963).
36. R. M. Peck, G. L. Miller, and H. J. Creech, *J. Am. Chem. Soc.*, **75**, 2364 (1953).
37. R. M. Herriott, *J. Gen. Physiol.*, **19**, 283 (1935–1936).
38. H. Buttkus, *J. Food Sci.*, **32**, 432 (1967).
39. H. Fraenkel-Conrat and H. S. Olcott, *J. Biol. Chem.*, **161**, 259 (1945).
40. H. Tamaoki, Y. Murase, S. Minato, and K. Nakanishi, *J. Biochem. (Tokyo)*, **62**, 7 (1967).
41. A. Matsushima, Y. Hachimori, Y. Inada, and K. Shibata, *J. Biochem. (Tokyo)*, **61**, 328 (1966).
42. E. Slobodian, G. Mechanic, and M. Levy, *Science*, **135**, 441 (1962).
43. B. S. Hartley, and B. A. Kilby, *Biochem. J.*, **56**, 288 (1954); J. R. Clark and L. Cunningham, *Biochemistry*, **4**, 2637 (1965).
44. A. Previero, M. A. Coletti, and L. Galzigna, *Biochem. Biophys. Res. Commun.*, **16**, 195 (1964).
45. S. Moore, *J. Biol. Chem.*, **238**, 235 (1963).
46. Y. Hachimori, H. Horinishi, K. Kurihara, and K. Shibata, *Biochim. Biophys. Acta*, **93**, 346 (1964).
47. J. B. Mudd, *J. Biol. Chem.*, **241**, 4077 (1966).
48. P. Edman, *Acta Chem. Scand.*, **4**, 277, 283 (1956); S. Eriksson and J. Sjoquist, *Biochim. Biophys. Acta*, **45**, 290 (1960).
49. L. Weil, A. R. Buchert, and J. Maher, *Arch. Biochem. Biophys.*, **40**, 245 (1952).
50. H. Fraenkel-Conrat, *J. Biol. Chem.*, **154**, 227 (1944).
51. J. F. Riordan, M. Sokolovsky, and B. L. Vallee, *J. Am. Chem. Soc.*, **88**, 4104 (1966).
52. T. Okuyama and K. Satake, *J. Biochem. (Tokyo)*, **47**, 454 (1960).

XII) bear some noticeable resemblance to the substrate **X**. It was found that a rapid stoichiometric reaction occurred with either compound, introducing one mole of reagent per mole of pepsin and leading to 100% loss of peptic activity. Chemical analysis subsequently showed that, indeed, β-carboxyl groups of aspartyl residues had been modified during these reactions.

The enzymes trypsin and chymotrypsin are known to be quite similar mechanistically and chemically. Their main differences are manifested in their differing specificities, since trypsin is highly specific for cationic substrates (e.g., derivatives of lysine and arginine), whereas chymotrypsin is highly specific for aromatic substrates (derivatives of tyrosine and tryptophan). Thus it is not surprising that the identification of their catalytically active histidine residues involved the reagents **XIIIa**, tosyl-L-lysine chloromethyl ketone, for trypsin[17] and **XIVa**, tosyl-L-phenylalanyl chloromethyl ketone, for chymotrypsin.[18] The corresponding substrates are shown as **XIIIb** and **XIVb**. Reagents **XIIIa** and **XIVa** react rapidly and stoichiometrically with trypsin and chymotrypsin, respectively, leading to 100% inactivation in

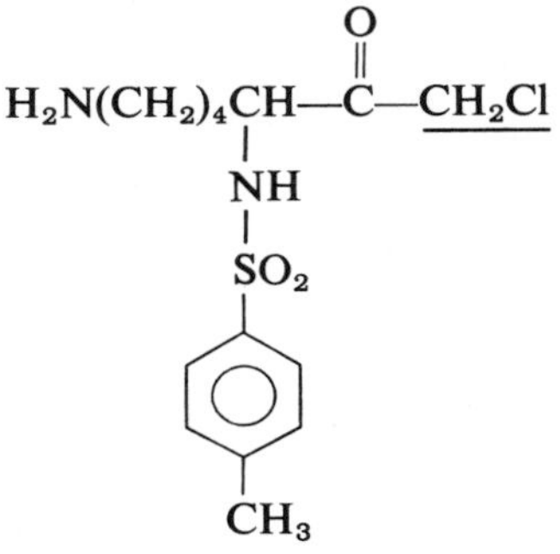

XIIIa

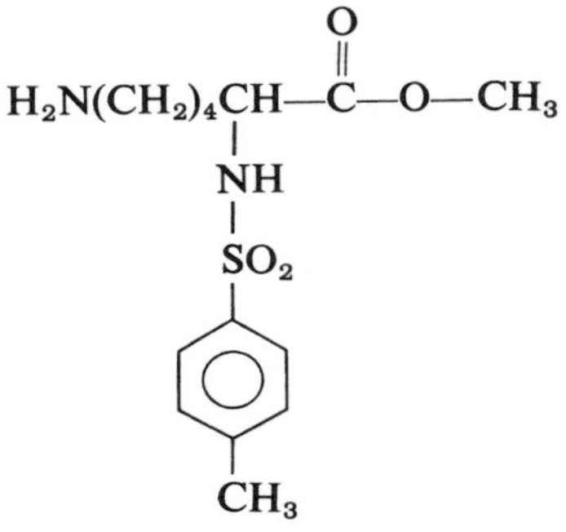

XIIIb

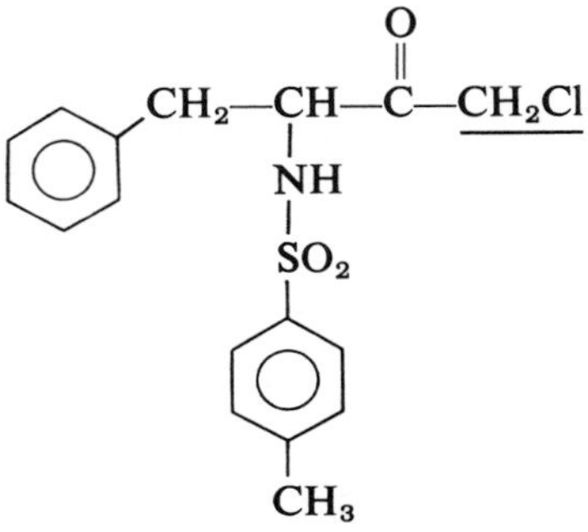

XIVa

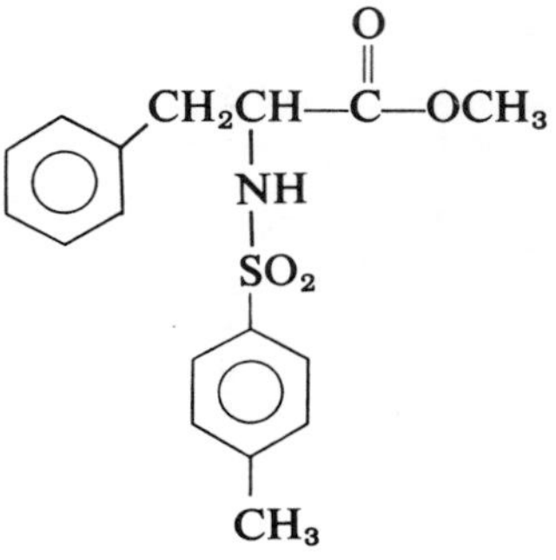

XIVb

both cases. In both cases the N-3 atom of the imidazole ring of a histidine residue is alkylated, with the concomitant displacement of chloride. This type of alkylation was discussed earlier in the chapter.

In both the carboxyl modifications of pepsin and the histidine modifications of chymotrypsin and trypsin the reactions that caused the inactivation were not typical of the type usually catalyzed by the enzymes involved. The pepsin case involved a transition metal ion catalyzed carbenoid alkylation of a carboxyl group, and the chymotrypsin and trypsin cases involved a nucleophilic displacement of a reactive halogen by an imidazole nitrogen. All three enzymes are hydrolases. Thus it is readily apparent that once the initial reaction had occurred, the enzyme was not equipped to complete its catalytic sequence of steps. Anthropomorphically speaking, the enzyme had been seduced into doing something it normally is not required to do and had been inactivated in the process.

The other approach to covalent labeling of the active site allows the enzyme to perform its normal function on a type of molecule it normally does not encounter. This leads to a covalent intermediate that the enzyme is again not equipped to handle. The classic example here is the use of phosphoryl and sulfonyl derivatives to covalently label the serine residue of the class of hydrolases now routinely referred to as serine hydrolases. With this class of enzymes the covalent labeling compounds are materials which are phosphorous or sulfur analogs of compounds that would quite likely be nonspecific substrates of the enzyme were the heteroatom replaced by the carbon atom.

The two most widely used covalent labels for these serine hydrolases have been compounds **II** and **III**. Diisopropylfluorophosphate (DFP), **II**, is particularly noteworthy, since the discovery of its reaction with acetylcholine esterase over two decades ago opened the door to the field of covalent labeling studies. Through the use of ^{32}P labeling techniques it was determined that one serine residue per molecule of enzyme had been phosphorylated and the derivatized enzyme was completely inactive, since the enzyme is incapable of catalyzing the dephosphorylation reaction.

Similarly, phenylmethanesulfonyl fluoride (PMSF), **III**, reacts with active serine residues in serine hydrolases to form inactive sulfonyl ester derivatives of the enzymes. A large variety of sulfonyl fluorides and fluorophosphates have been synthesized over the years in attempts to take advantage of a particular enzyme's specificity restrictions in the design of a suitable covalent label.

The coenzyme pyridoxal phosphate, **XV**, might be considered an unusual kind of covalent label, since in a large number of enzymes this coenzyme is found covalently attached to the ε-amino group of a lysine residue through a Schiff base linkage with the aldehyde group. Thus, when an enzyme is de-

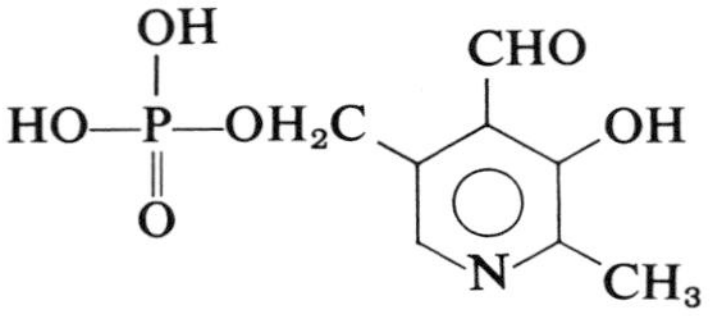

$$
\begin{array}{c}
\text{OH} \\
| \\
\text{HO—P—OH}_2\text{C} \\
|| \\
\text{O}
\end{array}
\quad
\begin{array}{c}
\text{CHO} \\
\text{OH} \\
\text{CH}_3
\end{array}
$$

XV

termined to have a pyridoxal phosphate requirement, one must suspect the involvement of a lysine residue. However, such lysine residues are not "catalytically active" groups mentioned previously. This is apparent from the reaction catalyzed by many pyridoxal phosphate enzymes (Eq. 3-11), where it is seen that the pyridoxal itself is the active "residue" in a transimination reaction. This reaction and other pyridoxal reactions will be discussed more fully in Chapter 8 on coenzymes.

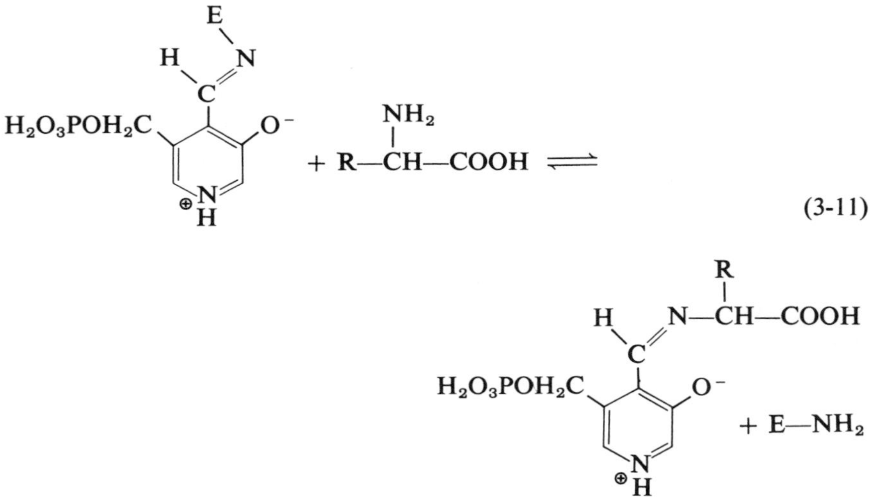

$$
\text{H}_2\text{O}_3\text{POH}_2\text{C} \underset{\text{pyridine}}{\overset{\text{C}=\text{N}-\text{E, O}^-}{}} + \text{R—CH(NH}_2)\text{—COOH} \rightleftharpoons
$$

(3-11)

$$
\text{H}_2\text{O}_3\text{POH}_2\text{C} \underset{\text{pyridine}}{\overset{\text{C}=\text{N—CH(R)—COOH, O}^-}{}} + \text{E—NH}_2
$$

There are many sulfhydryl reagents for general modification studies and some of these, notably PCMB, N-ethylmaleimide, and idoacetic acid or iodoacetamide, have been found to react especially rapidly with sulfhydryl-active enzymes. However, with sufficiently vigorous conditions, these reagents will react with any available sulfhydryl group, and the haloacetate derivatives have also been known to alkylate histidine residues.

Although methionine, tyrosine, and arginine all have reactive functional groups, and although the first two of these residues have been claimed to be actively involved in one or another enzyme's catalytic process, there is no firmly documented case where the specific reaction of one or two of these residues with a particular reagent has led to a 100% loss of catalytic activity. This is not meant to imply that the modification of such residues is not useful; indeed the first part of this chapter points up the great utility of general modification studies. Rather, it is mentioned in order to emphasize that the only amino acid residues that have been shown conclusively to be directly involved in any enzyme's catalytic process through covalent labeling techniques are serine, histidine, aspartic acid, cysteine, and lysine.* This may be due in part to the fact that too few enzymes have been studied in sufficient detail to allow the exact identification of their catalytic groups (particularly the metalloenzymes, where much elegant kinetic work has led to painfully slow advancement of our mechanistic understanding). A large enough number has been studied, however, to make it abundantly clear that the evolution of catalytic function must have been a remarkably conservative process. For all the wealth of serine enzymes in nature, there has yet to be discovered an enzyme with a catalytically active threonine.

Titrations of Enzyme Active Sites[19]

One of the most widely used techniques in general analytical chemistry is titration. The development of covalent active site labels for enzymes has also led to the development of titration methods for enzyme active sites. Such methods are conceptually the same as other titrations in that one has to determine a species to be measured, a satisfactory method for the quantitative measurement of that species, and a framework of theory to translate experimental measurements into molar concentrations.

For enzymes, where purity is often a problem, these requirements may mean designing the titrating reagent in accord with the specificity requirements of the enzyme and insuring that the titrant react with the enzyme only in its native conformation. The titrating reagent should be pure and well-characterized chemically and its reaction with the enzyme should lead to an "easily and accurately measured change in a physical property"[19]— most commonly a spectral change. Ideally, it will react in a stoichiometric, irreversible, covalent manner.

Kézdy and Kaizer[19] have listed the criteria that must be satisfied for a reagent to be a useful active-site titrant: (1) a rate constant comparison of the titrant's reactivity with the enzyme and with a suitable model system to determine the occurrence of catalysis in the enzymic reaction; (2) a kinetic

---

* Lysine involvement in aldolases has been demonstrated by trapping the intermediate Schiff base formed between enzyme and substrate by borohydride reduction and not by 100% inhibition through covalent attachment of an inhibitor. (See Chapter 11.)

study, which for an irreversible covalent titrant should ideally reveal first or second-order kinetics or exhibit saturation effects (see Chapter 4); competitive inhibitors (see Chapter 6) should inhibit the titration reaction—this is a good indication (but not a rigorous proof) of reaction at the active site; (3) the product of the irreversible one-to-one reaction between enzyme and titrant should be enzymically inactive, and if there are means to reverse the reaction, activity should be fully restored; (4) if the titrant is dissymmetric, the system should exhibit high optical specificity, particularly if the titrant is structurally analogous to a substrate for the enzyme; and (5) denaturation of the enzyme, for example, by heat, urea, or extreme of pH, should lead to the abolition of the titration reaction.

One limitation of this approach to determining enzyme active site molarities is the generally high concentrations of enzyme required for the observation of the change in a measured parameter caused by the reaction of enzyme and titrant. This is often as high as $10^{-5}M$ and routinely $10^{-6}M$. Rate assays may be useful down to $10^{-9}M$ in enzyme, but with rate assays one usually expresses concentration of enzyme in activity units per microgram of material used. For the determination of the absolute values of the rate constants for an enzymic reaction one must have some measure of the molar concentration of active enzyme, and the titration methods provide the most accurate measure of this. When the level of purity of an enzyme is known a titration study can also lead to an accurate measure of its molecular weight.

The first clear example of the use of a titration method to determine the molar concentration of an enzyme in solution was the work of Schonbaum et al.,[20] where the reagent cinnamoyl imidazole (**XVI**) was used to titrate the enzyme chymotrypsin by the reaction shown in Eq. 3-12. The change in extinction coefficient ($\Delta\varepsilon$) at 315 nm is $-8950M^{-1}\,cm^{-1}$ at pH 5.0. As long as the pH is maintained at 5 for this reaction, an essentially irreversible,

$$(3\text{-}12)$$

$$\Delta\varepsilon_{pH\,5.0}^{315} = -8950$$

XVI

stoichiometric, covalent active site modification obtains between chymotrypsin and **XVI**. With the spectrally measured change in absorbance and the known value of $\Delta\varepsilon_{315}$, one merely uses Beer's law ($\Delta OD = \Delta\varepsilon cl$) to determine c, the molar concentration of enzyme in the spectrophotometer cell. Since a known aliquot of enzyme is used, it is an easy matter to calculate the molar concentration of enzyme in the stock solution. Structure **XVI** is now widely used to titrate a variety of proteolytic enzymes similar to chymotrypsin.

Kézdy and Kaiser[19] also give details of more sophisticated titration methods which are often of value when an irreversible stoichiometric titrant is not available or cannot be found.

Reporter Groups

A reporter group is a group on a molecule which, after reaction with a protein, is able to "report" to the observer something about the nature of the environment where it is bound. This reporting is generally done spectrally, either by ultraviolet, electron paramagnetic resonance, nuclear magnetic resonance, or fluorescence spectroscopy.

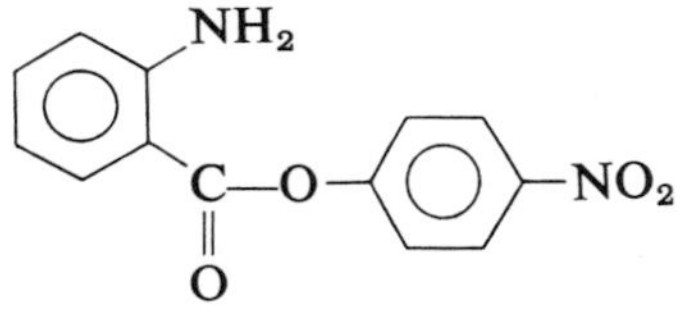

XVII

For example, Haughland and Stryer[21] have used the molecule p-nitrophenyl anthranilate (**XVII**) to introduce the fluorescent chromophore of anthranilic acid into the active site of the chymotrypsin molecule and then studied the fluorescent behavior of the modified enzyme.

Similarly, Koshland[22] has developed an ultraviolet probe to study the microenvironment of tryptophan residues in proteins. The reagent, 2-hydroxy-5-nitrobenzyl bromide (**XVIII**), alkylates only tryptophan residues and intro-

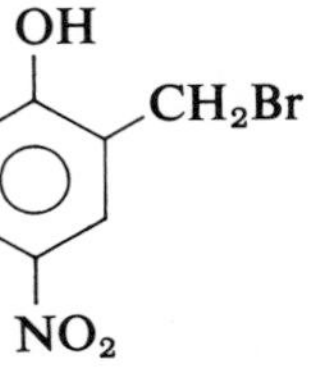

XVIII

duces a nitrophenol chromophore into the protein. Since most proteins are transparent at the λ_{max} of this chromophore (410 nm when ionized, 315 nm when associated), it provides a useful handle for study.

As is evident, this is a potentially powerful approach for learning about the anatomy of an enzyme and is limited mainly by the imagination of the investigator.

Addendum: Since completion of this text, two books have appeared providing much useful information on mechanisms and procedures for many of the chemical modifications discussed in this chapter.[23,24]

References

1. L. A. Cohen, *Ann. Rev. Biochem.*, **37**, 695 (1968).
2. A. N. Glazer, *Ibid.*, **39**, 717 (1970).
3. B. L. Vallee and J. F. Riordan, *Ibid.*, **38**, 733 (1969).
4. S. J. Singer, *Adv. Protein Chem.*, **22**, 1 (1967).
5. D. E. Koshland, Jr., *Adv. Enzymol.*, **22**, 45 (1960).
6. B. R. Baker, *Design of Active Site Directed Irreversible Enzyme Inhibitors*, Wiley, New York, 1967.
7. I. Svensson, *Compt. Rend. Trav. Lab. Carlsberg*, **19**, 347 (1968).
8. F. Sanger, The Structure of Insulin, in *Biochemical Research*, D. E. Green, Ed., Interscience, New York, 1956.
9. D. S. Sigman et al., *Biochemistry*, **8**, 4560 (1969).
10. J. F. Riordan, M. Sokolovsky, and B. L. Vallee, *Biochemistry*, **6**, 3609 (1967).
11. M. Sokolovsky, J. F. Riordan, and B. L. Vallee, *Biochemistry*, **5**, 3582 (1966).
12. T. C. Bruice, M. J. Gregory, and S. L. Walters, *J. Am. Chem. Soc.*, **90**, 1612 (1968).
13. S. L. Walters and T. C. Bruice, *J. Am. Chem. Soc.*, **93**, 2269 (1971).
14. W. J. Ray, Jr., *Meth. Enzymol.*, **XI**, 490 (1967).
15. W. W. Cleland, *Biochemistry*, **3**, 480 (1964).
16. L. E. Baker, *J. Biol. Chem.*, **193**, 809 (1951).
17. M. Mares-Guia and E. Shaw, *Fed. Proc.*, **22**, 528 (1963).
18. G. Schoellman and E. Shaw, *Biochemistry*, **2**, 252 (1963).
19. F. J. Kézdy and E. T. Kaiser, *Meth. Enzymol.*, **XIX**, 1 (1970).
20. G. R. Schonbaum, B. Zerner, and M. L. Bender, *J. Biol. Chem.*, **236**, 2930 (1961).
21. R. P. Haugland and L. Stryer, in *Conference on Biopolymers*, Academic, New York, 1968, p. 231.
22. D. E. Koshland, Jr., Y. D. Karkhanis, and H. G. Latham, *J. Am. Chem. Soc.*, **86**, 1448 (1964).
23. C. H. W. Hirs and S. N. Timasheff (eds.), *Meth. Enzymol.*, **XXV**, (1972)
24. G. E. Means and R. E. Feeny, *Chemical Modification of Proteins*, Holden-Day, San Francisco, 1971.

CHAPTER 4

Kinetics I

The Role of Kinetics in Enzyme Studies

Anyone who is concerned with enzymes is concerned with the general area of catalysis; and anyone who is concerned with catalysis is most certainly concerned with the velocities of chemical reactions, that is, with the area of chemical kinetics.

At the simplest level, an enzyme investigator must develop a so-called *assay* for an enzyme in order to monitor its purity and its integrity as a native (undenatured) enzyme. This assay involves adding a known amount of enzyme preparation to a solution containing a known concentration of substrate species and recording the rate at which substrate is converted to product under carefully controlled conditions of temperature, pH, ionic strength, and so on. The rate measurement allows the investigator to assign his enzyme preparation a *specific activity*, which is expressed as units of enzyme activity per milligram of protein. One unit (U) of any enzyme is defined by the Commission on Enzymes of the International Union of Biochemistry as that amount which will catalyze the transformation of one micromole of substrate per minute under defined optimal conditions. No small amount of effort must be expended by the investigator of a new enzyme preparation in the establishment of these optimal conditions for an assay and applying it in reproducible fashion through the several steps of isolation and purification discussed in Chapter 2.

A really adequate assay is capable of revealing *quantitatively* the amount of enzyme in a given solution or extract in terms of the catalytic effect the enzyme produces. Before such an ideal assay can be designed, however, one must know (1) the reaction catalyzed in stoichiometric detail, (2) what species beside the substrate itself must be present (metal ions, coenzymes, etc.), (3) the kinetic dependence of the reaction on each required species, (4) the optimum conditions of temperature, pH, and ionic strength, and (5) a suitable means for monitoring substrate disappearance or product appearance.

Beyond this relatively trivial role of kinetics in providing a criterion by which the purity and gross catalytic efficiency of an enzyme preparation can be gauged, any more fundamental study of the actual mechanism of catalysis must be based on careful quantitative measurements of the rate of the catalyzed reaction as conditions of enzyme concentration, substrate concentration, pH, temperature, ionic strength, and other factors are systematically varied. A great deal can be learned about the mechanism of an enzyme-catalyzed reaction from kinetic studies alone, and in fact kinetics is the only generally useful tool for the elucidation of enzyme mechanisms for all but the most highly purified enzyme preparations. Ideally, however, the kinetic studies should be accompanied by parallel investigations using nonkinetic techniques such as chemical modification (Chapter 3) and structural characterization (Chapter 12) which can be applied in the study of pure proteins.

Chemical Kinetics

Before we proceed to the specific application of kinetics to enzyme-catalyzed reactions, some of the basic fundamentals and terminology of chemical kinetics must be outlined. In general, chemical reactions may be classified at the *molecular* level in terms of how many molecules must interact to form the reaction products. Hence we have reactions which are *unimolecular*, *bimolecular*, and, much more seldom, *termolecular* or higher. (Simultaneous three-body collisions of moving particles are no more probable at the molecular level than they are on a pool table.)

Although perhaps useful conceptually, molecularity is not a good working classification for chemical reactions because the chemist cannot observe individual events at the molecular level. Chemical reactions can also be classified, purely on experimental grounds, in terms of *reaction order*. Here we enter the realm of chemical kinetics. For example, consider a hypothetical reaction where one mole of compound A reacts with one mole of compound B to give stoichiometrically a mole of product P:

$$A + B \rightarrow P \tag{4-1}$$

Now our first thought is to view this as a bimolecular reaction involving the simple collision of one A with one B molecule for every P formed. This may indeed be the case, and if so the *rate law* actually followed by the reaction (which can be derived from experimentally obtainable data) is found to be

$$\text{rate} = v = -\frac{d[A]}{dt} = -\frac{d[B]}{dt} = \frac{d[P]}{dt} = k[A][B] \tag{4-2}$$

where the brackets indicate molar concentrations and t is time (usually minutes or seconds). Equation 4-2 is a *second-order* rate law since the rate is directly proportional to the product of the concentrations of two reactants; or in other words, the reaction is *first order* in [A], first order in [B], and second order overall. The second-order rate constant k is simply the constant of proportionality between the reaction rate and the product of reactant concentrations. It has units of $M^{-1} \cdot \text{time}^{-1}$, consistent with the units of reaction rate, $M \cdot \text{time}^{-1}$.*

But Eq. 4-1 may describe a reaction that is not second order. It may turn out, for example, that the experimentally measured rate is directly proportional to [A] but has no dependence whatever on [B]. If so, our rate law is Eq. 4-3. This is a *first-order* rate law where the velocity of the reaction is

$$v = -\frac{d[\text{A}]}{dt} = -\frac{d[\text{B}]}{dt} = \frac{d[\text{P}]}{dt} = k[\text{A}] \tag{4-3}$$

proportional to a single reactant concentration and the first order rate constant, k, now has the units of $\sec^{-1}$ or $\min^{-1}$. Clearly in this case, the rate law does not reflect the stoichiometry of Eq. 4-1. This means that among the mechanisms that can be discarded are any that call for a rate-determining bimolecular collision of A and B. One mechanism that *is* consistent with both Eq. 4-1 and Eq. 4-3 is given in Eqs. 4-4 and 4-5.

$$\text{A} \xrightarrow{\text{slow}} \text{X} \tag{4-4}$$

$$\text{X} + \text{B} \xrightarrow{\text{fast}} \text{P} \tag{4-5}$$

The slow unimolecular transformation of A into *reactive intermediate* X determines the rate of the overall reaction. It is the so called *rate-determining* or *rate-limiting* step. The second bimolecular step is so fast relative to the first that it has no significant influence on the overall rate. Thus we find that this mechanism implies a reaction which is first order in [A] and *zero order* in [B].

We might note here that Eqs. 4-4 and 4-5 are by no means the only possible mechanism consistent with first-order kinetics and the stoichiometry of Eq. 4-1. Equation 4-6, for example, is just as plausible although not as appealingly simple.

$$\text{A} \xrightarrow[\text{slow}]{} \text{X} \xrightarrow[\text{fast}]{} \text{Y} \xrightarrow[\text{fast}]{+\text{B}} \text{P} \tag{4-6}$$

It is important to realize that kinetic studies alone never lead to a *unique* interpretation. We cannot write down *the* mechanism. Kinetic studies give us a rate law which, in conjunction with the overall stoichiometry of the reaction at hand, allows us to propose any number of possible mechanisms

* The units on the left side of an equation and on the right side must be the same. Since units of velocity = moles/liter/sec = $M \cdot \sec^{-1}$, and since [A] and [B] units = moles/liter = M, dimensional treatment of Eq. 4-2 shows that $(M \cdot \sec^{-1}) = k(M)^2$. Therefore the units of k (second order) are given by $M \cdot \sec^{-1}/M^2 = M^{-1} \cdot \sec^{-1}$.

that are consistent with the data. Further narrowing of the possibilities requires experiments of other kinds (e.g., detection and characterization of a reactive intermediate, systematic variation of reactant structure, variation of reaction medium and conditions, studies with isotopic labels). However, it is never possible to consider a mechanism *proved*, even when seemingly exhaustive experimental evidence of every conceivable variety appears to be subject to a unique mechanistic interpretation. Mechanisms of chemical reactions, be they enzyme-catalyzed or not, are elusive and can generally only be more or less tightly encircled, not definitely pinned down. Nonetheless, it is probably safe to say that kinetic studies are in general the most powerful and widely applicable *single* means of limiting the possibilities in a mechanism determination once the actual participants in the reaction (reactants, products, catalysts) have been identified.

a. Experimental Kinetics—The Determination of Reaction Order

Now that we have some appreciation for the place of kinetic studies in mechanism determinations, let us deal briefly with some of the techniques an investigator uses to obtain useful kinetic information.

In the general case of a reaction where n molecules must interact with each other in the rate-limiting step, we have

$$q\text{A} + r\text{B} + \cdots \xrightarrow{k} m\text{P} \tag{4-7}$$

$$v = -\frac{d[\text{A}]}{q\,dt} = -\frac{d[\text{B}]}{r\,dt} = \frac{d[\text{P}]}{m\,dt} = k[\text{A}]^q[\text{B}]^r \ldots, \qquad n = q + r + \cdots \tag{4-8}$$

where q, r, ..., are individual orders in $[\text{A}]$, $[\text{B}]$, ..., and n is the overall order of the reaction, and k has the dimensions of $M^{1-n}\cdot\text{time}^{-1}$. (Note carefully here that Eq. 4-7 is the equation for the *single* rate-limiting step in a reaction mechanism.) This general case in itself need not concern us too much, since, as stated earlier, there are few reactions involving collisions of more than two molecules simultaneously. Reactions with three or more reactant molecules usually proceed stepwise with bimolecular or unimolecular steps. Such reactions may follow first-, second-, third-, or mixed-order kinetics depending on which of the steps is rate limiting. Even fractional reaction orders are not uncommon.

There are a number of ways to determine the order of a reaction.[1] Perhaps the most generally useful and dependable method is to vary the concentration of each reactant individually while holding all other concentrations and conditions constant in order to determine the effect of each reactant on the rate. The individual orders in each reactant are obtained as the slopes of plots of log v versus log [reactant]. This may be readily appreciated in Eq.

4-9 where we have simply taken the logarithm of each side of Eq. 4-8. In

$$\log v = \log k + q \log [A] + r \log [B] + \cdots \qquad (4\text{-}9)$$

such a plot, the only variables are v and one concentration term. All other terms remain constant. Thus, for example, if we measured the initial rate of the reaction with six different initial concentrations of A while holding all other reactant concentrations the same, and then plotted our six values for $\log v_0$ against $\log [A]_0$ (where the subscript zero indicates initial value), the plot would be a straight line with slope q and intercept ($\log k + r \log [B]_{const} + \cdots$). The values for v_0 needed to make such a plot are obtained as indicated in Figure 4-1.

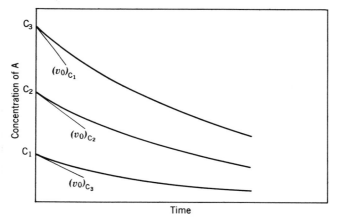

Time

FIGURE 4-1. Progress curves for change in concentration of A as a function of time at three different initial concentrations of A (C_1, C_2, C_3). The slope of the tangent at time = 0 for each curve is equal to the initial rate v_0 for that curve.

Aside from this very general method for determining reaction order, one can sometimes test a single progress curve to see if it fits the correct mathematical form for a given rate law. For this purpose, the differential forms of the possible rate laws (such as Eq. 4-2 or 4-3) must be integrated to give a graphable expression. The appropriate integrated rate expressions in slope-intercept form for the first-order (Eq. 4-3) and second-order (Eq. 4-2) rate laws are given as Eqs. 4-10 and 4-11, respectively. The subscript t means at time t. The X in Eq. 4-11 is the amount of A or B which has reacted at

$$\ln [A]_t = \ln [A]_0 - kt \qquad (4\text{-}10)$$

$$\ln \left(\frac{[B]_0 - X}{[A]_0 - X} \right) = \ln \frac{[B]_0}{[A]_0} + ([B]_0 - [A]_0)kt \qquad (4\text{-}11)$$

time t (or, equivalently, $X = [P]_t$). Another useful (and much simpler) integrated second-order rate expression holds for the special case where $[A]_0 = [B]_0$. In that case if we let c equal the remaining concentration of A or B at any time t, then

$$v = -\frac{dc}{dt} = kc^2 \tag{4-12}$$

and integration gives

$$\frac{1}{c} = \frac{1}{c_0} + kt \tag{4-13}$$

It may be readily seen that plotting $\ln [A]_t$ against time will give a straight line (whose slope is k) if the reaction is first order. A second-order reaction will give a straight line if $\ln \{([B]_0 - X)/([A]_0 - X)\}$ (or in the special case $1/c$) is plotted against time. In similar fashion, any differential rate law may be integrated to give a graphable time-dependent function of concentration which may be tested for fit against the data from a single progress curve.

Great care and caution must be exercised in the use of integrated rate law plots to determine reaction orders for single progress curves. There will *always* be some experimental error in the data, and small but important deviations from linearity can very easily be obscured by this random error of the measurements. This is especially true if the data are taken only over a small part of the extent of reaction, say the first 5 to 10%, where the rate may be very nearly constant and the data can be deceptively well fitted by *any* of the simple integrated equations. The method requires fewer experiments than others, but the conclusions are often less trustworthy.

As a third alternative, one may choose to evaluate reaction order in terms of the measured reaction *half-time*. A reaction half-time is the time required for the concentration of a reactant to fall to exactly one-half its initial value or for the concentration of a product to build up to one-half its final value. Consider, for example, the integrated rate law for a first order reaction (Eq. 4-10). If we substitute $[A]_0/2$ for $[A]_t$ and $\tau_{1/2}$ (half-time) for t in that equation, we arrive at an expression (Eq. 4-14) giving the half-time $\tau_{1/2}$ as a very simple function of k for first order reactions:

$$\ln \frac{[A]_0}{2} = \ln [A]_0 - k\tau_{1/2}$$

$$k\tau_{1/2} = \ln [A]_0 + \ln 2 - \ln [A]_0 = \ln 2 \tag{4-14}$$

$$\tau_{1/2} = \frac{\ln 2}{k}$$

Similar treatment of Eqs. 4-11 and 4-13 give half-time Eqs. 4-15 and 4-16 for second-order reactions:

$$\tau_{1/2} = \frac{1}{k([B]_0 - [A]_0)} \ln \frac{2[B]_0 - [A]_0}{[B]_0} \qquad (4\text{-}15)$$

$$\tau_{1/2} = \frac{1}{k[c]_0} \qquad (4\text{-}16)$$

The especially noteworthy feature of these expressions is that first-order half-times are (uniquely) independent of any concentration terms, whereas the half-times for higher or lower (fractional) order reactions have concentration dependences. Thus first-order reactions are easily identified as such by observing a constant half-time for the reaction in several experiments with widely different initial reactant concentrations.

It will be apparent at this point that first-order reactions are much easier to deal with than any other common reactions. Equation 4-14 enables us to derive first-order rate constants without ever determining the absolute concentrations of any reactant or product. Thus instead of plotting ln [A] against time, one can plot directly the time dependence of any experimentally measurable quantity which is proportional to the concentration of A (or P) without knowing the constant of proportionality (see, for example, Figure 4-2). Such measurable quantities include absorbance changes (from ultraviolet, visible, infrared, or NMR spectroscopy) as well as changes in optical

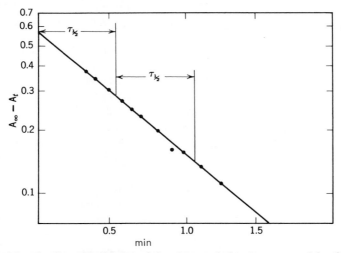

FIGURE 4-2. Semilogarithmic plot of the change of absorbance caused by the release of a chromophoric product in a first-order reaction. Note that the half-time, $\tau_{1/2}$, is independent of the extent of reaction.

rotation, refractive index, conductivity, volume, release of heat, or volume of titrant required to react with some species generated in the reaction. The only requirement is that the measured quantity has a linear relationship to the concentration of a reactant or a product (or indeed even to the concentrations of several species present in the reaction mixture combined).[2]

Since first-order kinetics are so readily dealt with, kineticists often *force* their more complex reactions to proceed in *pseudo-first-order* fashion. This requires that all of the reactants except one be present initially in large excess. As long as the order of the reaction with respect to that one single reactant is first order, the reaction will yield first-order kinetics as a result of the fact that the concentrations of all reactants present in excess will remain virtually constant as the reaction proceeds. By allowing each reactant in turn to yield its own pseudo-first-order rate constant, the investigator is enabled to piece together the rate laws for exceedingly complex reactions.

b. Simple Rate Theory and Catalysis

Let us turn again to our simple reaction in which A and B react to produce P in a single bimolecular step. For every P produced in this case, there must be a collision of an A molecule with a B molecule. This much is self-evident. But it is rarely true that *every* A-B collision which occurs results in a reaction. In general only a small fraction of A-B collisions are both forceful enough and correctly aligned to produce a reaction. Consider, for example, the reaction that occurs between a chlorine atom and chloroform to produce HCl and a trichloromethyl radical (Eq. 4-17). The requirement for reaction here is that the chlorine atom hit the hydrogen atom of the chloroform rather than one

$$:\overset{..}{\underset{..}{Cl}}\cdot \;+\; H\!-\!CCl_3 \;\longrightarrow\; [:\overset{..}{\underset{..}{Cl}}\text{---}H\text{---}CCl_3]^{\ddagger} \;\longrightarrow\; :\overset{..}{\underset{..}{Cl}}\!-\!H \;+\; \cdot CCl_3 \qquad (4\text{-}17)$$

of the other atoms in the chloroform molecule, and that this collision be energetic enough so that when the *activated complex* ($\ddagger$) breaks up, it results in the products shown rather than in unchanged reactants. In other words, Cl—H bond making and H—C bond breaking will not occur unless the collision has the correct alignment and unless the collision occurs with greater than a certain minimum amount of energy required to realign the bonding electrons in the new way. So of all the collisions that occur in, say, one second, only a small fraction of the total will have what it takes to produce a reaction. Most of the collisions will either be too weak or incorrectly aligned. In general, any reaction has its own unique requirements for effective collisions which depend on the structural characteristics of the reacting molecules and on the energy required to make and break the chemical bonds in question. This is why reactions have different reaction rates.

We can illustrate the situation we have just described by means of an

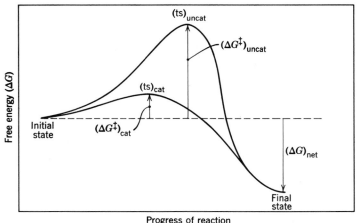

FIGURE 4-3. Energy diagram for a chemical reaction, uncatalyzed and catalyzed. (ts)$_{uncat}$ and (ts)$_{cat}$ are the transition states for the uncatalyzed and catalyzed reactions, respectively. $\Delta G^{\ddagger}$ indicates free energy of activation. $(\Delta G)_{net}$ is the overall free-energy change of the reaction.

energy diagram such as Figure 4-3, which also shows the role of a catalyst in a chemical reaction. The transition state is the energy-rich state of the interacting molecules (the activated complex, $\ddagger$, as in Eq. 4-17), which a collision must produce if a reaction is to occur. The actual minimum amount of energy needed to attain this transition state is called the *activation energy*,* indicated in the diagram by a vertical arrow. The rate of a reaction is proportional to the concentration of the transition-state species. A rise in temperature, because it increases the average thermal kinetic energy of the molecule population, increases the number of collisions capable of producing the transition state. Thus a rise of approximately 10°C in temperature will often double the reaction rate. A catalyst increases the reaction rate by actually lowering the activation energy barrier of the reaction. It does this by interacting with one or more reactants in such a way that less energy is required to achieve the necessary bond reorganizations associated with the reaction. And although the catalyst is very much involved in the reaction—it is included in the activated complex along with the reacting molecules—it is released unchanged when the reaction is completed. Enzymes are such catalysts.

---

* The activation energy quantity used here reflects both the collision force requirement and the collision alignment requirement for attaining the transition state; that is, both the activation enthalpy $\Delta H^{\ddagger}$ and the activation entropy $\Delta S^{\ddagger}$. This follows from the thermodynamic definition of free energy, $\Delta G = \Delta H - T\Delta S$.

Enzyme Kinetics

For most enzyme-catalyzed reactions, if one records the amount of substrate transformed (or product formed) as a function of time, the result is a progress curve such as that in Figure 4-4 in which velocity falls with time.

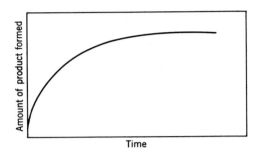

FIGURE 4-4.   Typical progress curve for an enzyme-catalyzed reaction.

Many factors may contribute to this drop in rate with time in addition to the expected falling off due to the decreasing concentration of unconverted substrate. For example, the product may inhibit the enzyme, or the enzyme may become denatured owing to its instability toward the conditions of temperature or pH employed, or the reverse reaction may become significant as the concentration of product builds up, or there may be a combination of these and other rate-influencing factors. Thus in all but the simplest of cases, attempts to derive equations describing progress curves for enzyme-catalyzed reactions are rarely made at an early stage of an investigation. The approach usually adopted in studying enzyme kinetics is to determine only *initial* velocities, varying only one initial condition at a time while all others are held constant. This avoids the complicating factors mentioned, because at the initial point they have not yet had time to operate, and all conditions are accurately known. However, mechanistic conclusions arising from these initial rate studies are markedly strengthened if the time course behavior predicted by these conclusions is indeed borne out experimentally.

Initial velocities are obtained experimentally as shown in Figure 4-1. It is necessary only to determine the first part of the progress curve (usually about the first 20%) and to obtain the slope of a tangent drawn at the origin of this curve which will equal v_0.

The factors that most commonly influence initial velocities of enzyme-catalyzed reactions are (1) initial enzyme concentration $[E]_0$, (2) initial substrate concentration $[S]_0$, (3) pH, (4) temperature, and (5) initial concentrations of other added species which may be activators, inhibitors, or cofactors. We will deal in this chapter with the first two of these factors as they influence simple one-substrate reactions, and in a general way with the temperature factor. Factors 3 and 5 will be considered in subsequent chapters.

The velocity of any single chemical transformation increases with temperature, approximately doubling for every increase of roughly 10°C. Experimentally, this is manifested in a temperature dependence of individual rate constants which takes the form

$$k = Ae^{-E^{\ddagger}/RT} \quad \text{or} \quad \ln k = -\frac{E^{\ddagger}}{RT} + \ln A$$

where A is known as the preexponential factor, $E^{\ddagger}$ is the so-called Arrhenius energy of activation, R is the gas constant, and T is the temperature in °K. Clearly, $\ln A$ and $E^{\ddagger}$ can be determined as the intercept and slope, respectively, of a linear plot of the dependence of $\ln k$ on $1/RT$. Except in the simplest of cases, these empirical quantities A and $E^{\ddagger}$ can be assigned no precise significance in terms of reaction rate theory. However with reference to the simple rate theory discussion earlier, the magnitude of A is to some extent a reflection of the collision orientation or alignment requirement of the reaction and $E^{\ddagger}$ reflects the collision force requirement. In this sense, values of A and $E^{\ddagger}$ may be of some use in determining reaction mechanisms.

Enzyme-catalyzed reactions usually involve several steps, each of which has its own temperature dependence. In addition, temperature affects such important factors as protein denaturation, protein ionization state, and the solubilities of species in solution. Thus the temperature dependence of an enzyme-catalyzed reaction may be quite difficult to interpret in a meaningful mechanistic sense. Nevertheless, the effect of temperature on enzyme reactions is sometimes determined and it may be of some significance to note that (as is the case for catalyzed reactions in general) enzyme-catalyzed reactions are less sensitive to temperature changes than their uncatalyzed counterparts. That is, whereas the uncatalyzed reaction rate may double with every 10° elevation of temperature, the enzyme-catalyzed reaction rate will increase by a factor somewhat less than two.

The reader who is interested in a more complete treatment of temperature effects on enzyme kinetics and in complex multisubstrate systems which are beyond the scope of this text is referred to the suggestions for further reading at the end of this chapter.[3–6]

a. Effect of Enzyme Concentration

At the beginning of this chapter, we noted that it is possible to develop quantitative enzyme assay procedures whereby the concentration of active enzyme present in a solution can be determined from the catalytic effect it produces. That is, the rate of enzyme-catalyzed conversion of substrate to product is, under assay conditions, directly proportional to the enzyme concentration (i.e., $v_0 = k[E]_0$ at constant assay conditions of substrate concentration, pH, temperature, etc.). This linear relationship between initial velocity and enzyme concentration (i.e., initial rate is first order in enzyme concentration) will in general be observed when the following conditions are met:

1. The method being employed to follow the reaction rate truly reflects the velocity of substrate-to-product conversion at all enzyme concentrations used.

2. There is no inhibitor or toxic impurity present in the reaction mixture that would inactivate some of the enzyme present.

3. The concentrations of substrate and any required cofactor are greatly in excess of the enzyme concentration.

b. Effect of Substrate Concentration

Almost invariably, for enzyme-catalyzed reactions involving a single substrate* which is present in large excess over the enzyme concentration, a plot of initial velocity v_0 against initial substrate concentration $[S]_0$ will give a section of a rectangular hyperbola as in Figure 4-5. This kinetic behavior led L. Michaelis and M. L. Menten in 1913 to a general theory accounting for enzyme action. This theory, as extended by G. E. Briggs and J. B. S. Haldane in 1925, is the basic foundation upon which virtually all of enzyme kinetics is built.

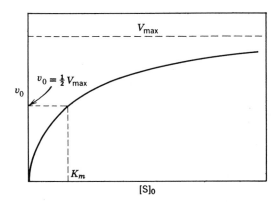

FIGURE 4-5. Hyperbolic form of typical substrate concentration curve.

* Also for multisubstrate reactions when all other substrate concentrations are held constant.

Michaelis-Menten theory holds that an enzyme E first reacts with substrate S reversibly to form an enzyme-substrate complex ES, which then breaks down in a second step to regenerate free enzyme and the product(s) P:

$$E + S \underset{k_{-1}}{\overset{k_1}{\rightleftharpoons}} ES \xrightarrow{k_{cat}} P + E \tag{4-18}$$

Let us see how this relatively simple mechanism gives rise to the behavior of Figure 4-5.

The following kinetic equations describe the rates of formation for each of the species in our proposed mechanism (Eq. 4-18):

$$\frac{d[S]}{dt} = -k_1[S][E] + k_{-1}[ES] \tag{4-19}$$

$$\frac{d[ES]}{dt} = k_1[S][E] - (k_{-1} + k_{cat})[ES] \tag{4-20}$$

$$\frac{d[E]}{dt} = (k_{-1} + k_{cat})[ES] - k_1[E][S] \tag{4-21}$$

$$\frac{d[P]}{dt} = k_{cat}[ES] \tag{4-22}$$

In addition, we have the *conservation* Eq. 4-23, which recognizes that the total enzyme concentration $[E]_0$ is partitioned between two species, free enzyme E and complexed enzyme ES. To be entirely rigorous, we should also

$$[E]_0 = [E] + [ES] \tag{4-23}$$

note that the total substrate concentration is similarly partitioned between S and ES. But we neglect this point since $[S] \gg [E]_0$ and thus $[ES]$ will always be insignificantly small compared to $[S]$ even if all of the enzyme is in the ES form.

The measurable rate of this reaction (Eq. 4-18) is simply the quantity $d[P]/dt$, the rate at which product is formed. However, Eq. 4-22 as it stands is of no use to us since it describes this rate as a function of $[ES]$, a quantity whose value we cannot measure experimentally. What we should like to do is derive from the given equations an expression for the reaction rate $d[P]/dt$ as a function of *measurable* concentrations. The only measurable concentrations here are the substrate concentration $[S]$ and the total enzyme concentration $[E]_0$. Unfortunately it is not possible to arrive at such an expression without first making simplifying assumptions.

The assumption that Michaelis and Menten originally made in 1913 was that $k_{cat} \ll k_{-1}$. This means that the first step (the reversible step) of Eq. 4-18 is a simple equilibrium which may be described in the usual fashion as

$$K_s = \frac{k_{-1}}{k_1} = \frac{[E][S]}{[ES]} \tag{4-24}$$

where K_s is just the dissociation constant for the enzyme-substrate complex. We can now proceed as follows toward our useful rate expression. Equation 4-23 provides a quantity which we can substitute for [E] in Eq. 4-24:

$$K_s = \frac{([E]_0 - [ES])[S]}{[ES]} = \frac{[E]_0[S]}{[ES]} - [S] \tag{4-25}$$

Solving Eq. 4-25 for [ES], we get

$$[ES] = \frac{[E]_0[S]}{K_s + [S]} \tag{4-26}$$

Now substituting for [ES] in Eq. 4-22, we have our final expression where the reaction rate is given in terms of constants and experimentally determinable concentration terms:

$$v = \frac{d[P]}{dt} = \frac{k_{cat}[E]_0[S]}{K_s + [S]} \tag{4-27}$$

In order to put our rate expression into final form, which will correspond mathematically to the hyperbola of Figure 4-5, we need to make one more substitution. Equation 4-22 tells us that the rate of the reaction is directly proportional to the concentration of the enzyme-substrate complex [ES]. Thus the absolute limit on the rate of the reaction will be reached when all of the available enzyme is in the ES form, that is when $[ES] = [E]_0$. Hence we can write

$$\text{maximum rate} = V_{max} = k_{cat}[E]_0 \tag{4-28}$$

Substituting in Eq. 4-27, then, and letting $K_s = K_m$ (see below) we arrive at the final well-known Michaelis equation:

$$v = \frac{V_{max}[S]}{K_m + [S]} \tag{4-29}$$

which can be recast into more familiar form as the equation for a rectangular hyperbola:

$$(V_{max} - v)(K_m + [S]) = V_{max}K_m \tag{4-30}$$

$$(a - y)(b + x) = \text{const}$$

It should be noted that when $[S] = K_m$, Eq. 4-29 simplifies to

$$v = \frac{V_{max}K_m}{K_m + K_m} = \frac{V_{max}}{2} \qquad (4\text{-}31)$$

The value of $[S]$ which is found *experimentally* to give half the maximal velocity is called the Michaelis constant K_m. The assumption we made earlier—that equilibrium is maintained between ES, E, and S—implies the equality of the experimental quantity K_m and the presumed equilibrium constant K_s. This assumption may or may not be valid in a given case. But whether or not K_m has any real significance as an equilibrium constant, the fact remains that for a large number of enzyme-catalyzed reactions, the two kinetic parameters K_m and V_{max} are both necessary and sufficient to define the rate law as indeed we have demonstrated in the correspondence between Eq. 4-29 and 4-30.

Whatever assumptions we make concerning the mechanism (Eq. 4-18), it will be true that the factors which influence the reaction rate produce their effects in two different ways. They may affect the rate at which the enzyme-substrate complex is formed (which is reflected in K_m) or the rate at which it breaks down to products (which is reflected in V_{max}). Thus it becomes a matter of some importance to determine V_{max} and K_m as accurately as possible. The hyperbolic plot of v_0 versus $[S]_0$ is not particularly useful for this purpose because it is not possible in general to determine precisely the asymptotic value V_{max} as $[S]_0 \rightarrow \infty$. The problem is avoided by recasting Eq. 4-29 into one of the following linear forms:

Double reciprocal form of Lineweaver and Burk (plot $1/v_0$ versus $1/[S]_0$):

$$\frac{1}{v_0} = \frac{K_m}{V_{max}} \cdot \frac{1}{[S]_0} + \frac{1}{V_{max}} \qquad (4\text{-}32)$$

Single reciprocal form of Eadie and Hofstee (plot $v_0/[S]_0$ versus v_0):

$$\frac{v_0}{[S]_0} = -\frac{v_0}{K_m} + \frac{V_{max}}{K_m} \qquad \text{or} \qquad v_0 = V_{max} - K_m \cdot \frac{v_0}{[S]_0} \qquad (4\text{-}33)$$

Figure 4-6 indicates the way in which these linear transforms of Eq. 4-29 may be employed to determine V_{max} and K_m.

Remember once again that although V_{max} has a well-defined meaning, both experimentally and theoretically in terms of our mechanism, as the maximum velocity approachable in the limit as enzyme approaches saturation by substrate, the same is not true for K_m. Experimentally, K_m is just the substrate concentration required for half maximal velocity under specified conditions:

$$K_m = [S]_{0(v_0 = V_{max}/2)} \qquad (4\text{-}34)$$

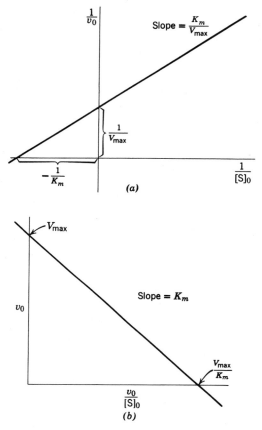

FIGURE 4-6. Plotting linear transforms of Michaelis-Menten equation. (a) Lineweaver-Burk plot. (b) Eadie-Hofstee plot.

But its theoretical significance depends upon the mechanism followed in a particular case, and many plausible mechanisms can be devised, all of which give rise to the identical rate law, Eq. 4-29. Here once again we emphasize the limitation of kinetics stated earlier—kinetic studies alone never lead us to a unique mechanistic interpretation for a given reaction. Kinetics allow us only to write a rate law that will be consistent with a large family of plausible mechanisms. Further narrowing of the possibilities requires other kinds of experiments.

Let us now reexamine the simple Michaelis-Menten mechanism and its kinetic equations (Eqs. 4-18 to 4-22). We have already noted that it is not possible to derive an explicit rate law as a function of determinable variables for this mechanism as written without making simplifying assumptions. We were able, following the equilibrium assumption of Michaelis and Menten, to derive a useful rate law. But the Michaelis-Menten equilibrium assumption is valid only in a very limited number of cases. It is far more generally true that k_{cat} in Eq. 4-18 is comparable to or greater than k_{-1}. Briggs and Haldane in 1925 formulated an alternative assumption which applies regardless of the relative magnitudes of k_{-1} and k_{cat}, but which still permits the derivation of an explicit rate law. The assumption is that as the reaction proceeds, at any given moment in time the rates of formation and breakdown of the ES complex are essentially equal. This is the so-called *steady-state* assumption, which is often invoked with respect to reactive intermediates postulated in reaction mechanisms. Mathematically, the steady-state approximation can be expressed in our case as

$$\frac{d[ES]}{dt} \simeq 0 \qquad (4\text{-}35)$$

Using this approximation, Eq. 4-20 becomes

$$\frac{d[ES]}{dt} \simeq 0 \simeq k_1[S][E] - (k_{-1} + k_{cat})[ES] \qquad (4\text{-}36)$$

Substituting the quantity $([E]_0 - [ES])$ for $[E]$ in Eq. 4-36 and solving for $[ES]$, we have

$$[ES] = \frac{k_1[S][E]_0}{k_1[S] + k_{-1} + k_{cat}} \qquad (4\text{-}37)$$

Finally, substituting this quantity for $[ES]$ in Eq. 4-22, we obtain

$$v = \frac{d[P]}{dt} = k_{cat}[ES] = \frac{k_{cat}k_1[S][E]_0}{k_1[S] + k_{-1} + k_{cat}} \qquad (4\text{-}38)$$

Rearranging Eq. 4-38 gives

$$v = \frac{k_{cat}[E]_0[S]}{(k_{-1} + k_{cat})/k_1 + [S]} = \frac{V_{max}[S]}{K_m + [S]} \qquad (4\text{-}39)$$

which is identical to Eq. 4-29. Thus the steady-state assumption leads to the same final rate equation that the equilibrium assumption of Michaelis and Menten led to, illustrating that the mechanism (Eq. 4-18) is consistent with the observed behavior of Figure 4-5 even when no constraints are placed on the relative magnitude of the rate constants.

It will be noted that the final result (Eq. 4-39) of the Briggs-Haldane treatment includes the Michaelis-Menten treatment as a special case. That is, when $k_{cat} \ll k_{-1}$, then $(k_{-1} + k_{cat})/k_1 \simeq k_{-1}/k_1$, which is just K_s.

Unfortunately, K_m and K_s are frequently used interchangeably. This of course is not generally valid, since K_m is an *experimentally meaningful* parameter and should not be regarded as the dissociation constant for the enzyme-substrate complex unless it has been specifically determined in a given case that this is so.

Moreover, it must be realized that most enzyme-catalyzed reactions are far more complex than the simple idealized case we have been discussing. For example, many such reactions may involve not one but several enzyme-substrate and/or enzyme-product complexes in sequence, as in Eq. 4-40 (see Chapter 5).

$$E + S \rightleftharpoons (ES)_1 \rightleftharpoons (ES)_2 \rightleftharpoons (EP) \rightleftharpoons E + P \qquad (4\text{-}40)$$

Yet such complexity is usually not apparent from the dependence of initial reaction rates on initial substrate concentration.

Other kinds of kinetic data may reveal the extent and nature of the complexity of a given enzyme-catalyzed reaction, but the starting point in the kinetic analysis of virtually all such reactions is the Michaelis-Menten relationship as derived above.

References

1. S. L. Friess, 'E. S. Lewis, and A. Weissberger (Eds.), *Technique of Organic Chemistry,* 2nd ed., Vol. VIII, Part I, Interscience, New York, 1963, Chap. 5.
2. W. P. Jencks, *Catalysis in Chemistry and Enzymology*, McGraw-Hill, New York, 1969, Chap. 11, pp. 559–560.
3. K. J. Laidler, *The Chemical Kinetics of Enzyme Action*, Oxford, Fairlawn, N.J., 1958.
4. C. Walter, *Steady-State Applications in Enzyme Kinetics*, Ronald, New York, 1965.
5. M. Dixon and E. C. Webb, *Enzymes*, 2nd ed., Academic, New York, 1964.
6. P. D. Boyer (Ed.), *The Enzymes*, 3rd ed., Vol. 2, Academic, New York, 1970.

CHAPTER 5

Kinetics II

Kinetics for the Acyl-Enzyme Case

Chapter 4 treated elementary enzyme phenomena as exemplified by the derivation of the Michaelis-Menten equation. This equation (5-1), which still is widely applied in many enzyme studies, was derived from the reaction

$$v = \frac{k_{cat}[E]_0[S]}{K_m + [S]} \tag{5-1}$$

scheme (Eq. 5-2), which implies rapid formation of a steady-state concentration of the intermediate ES complex. Beginning in the mid-1950s a number

$$E + S \underset{k_{-1}}{\overset{k_1}{\rightleftharpoons}} ES \overset{k_{cat}}{\longrightarrow} E + \text{Products} \tag{5-2}$$

of researchers studying proteolytic enzymes began to notice some unusual effects in the kinetics of the hydrolysis of certain substrates by these enzymes. For example, the k_{cat} values determined for the hydrolysis of a series of alkyl esters of N-acetyl-L-tryptophan by chymotrypsin were all identical. Another example was the behavior of p-nitrophenyl acetate with this enzyme, which showed a rapid initial release of nitrophenol followed by a slow, zero-order portion of the reaction.* The culmination of these studies came with a series of papers in 1964 by Bender[1] and co-workers, wherein the existence of another kinetically important intermediate after the ES complex was shown to be the minimum necessary expansion of Eq. 5-2. This intermediate, shown in Eq. 5-3 as ES′, is called the *acyl enzyme* for the serine

$$E + S \underset{k_{-1}}{\overset{k_1}{\rightleftharpoons}} ES \overset{k_2}{\longrightarrow} \underset{\underset{P_1}{+}}{ES'} \overset{k_3}{\longrightarrow} E + P_2 \tag{5-3}$$

and sulfhydryl hydrolases since the work done to date on these enzyme classes has shown that this ES′ is an ester derived from the acyl portion of the

---

* Observation of such behavior gave rise to the development of the enzyme titration methods, discussed in Chapter 3.

substrate and from the active hydroxyl or sulfhydryl group of the enzyme. Kinetic evidence has also been presented by Zeffren and Kaiser,[2] who suggest the involvement of an ES' species in pepsin catalysis. The exact nature of this intermediate is still under study; it seems clear from kinetic studies, however, that the functional group on the enzyme involved in ES' is a carboxyl group.[3] Since these ES' intermediates must exist during all reactions of these enzymes, and since normal Michaelis-Menten kinetics are generally observed for their reactions, the kinetic treatment that follows will be observed to lead to an expression that is experimentally indistinguishable from Eq. 5-1.

a. The Acyl-Enzyme Expression (Steady-State)

Let us assume that the following statements hold for the reacting system:

1. $[S]_0 \gg [E]_0$, that is, there is a large excess of substrate over enzyme.
2. The steady-state treatment conditions are valid.
3. $d[ES']/dt \ll d[S]/dt$.
4. $k_2[ES] = k_3[ES']$.

These assumptions will be used to treat the scheme of Eq. 5-3. Define a dissociation constant K_s as shown in Eq. 5-4 for the ES complex.

$$K_s = \frac{[E][S]}{[ES]} = \frac{k_{-1}}{k_1} \tag{5-4}$$

One may express $[E]_0$ in terms of $[E]$, $[ES]$, and $[ES']$, the species of enzyme present in solution:

$$[E]_0 = [E] + [ES] + [ES'] \tag{5-5}$$

From Eq. 5-4 we get

$$[E] = \frac{[ES]K_s}{[S]} \tag{5-6}$$

From assumption (4) we get

$$[ES'] = \frac{k_2}{k_3}[ES] \tag{5-7}$$

Substituting for E and ES' in Eq. 5-5,

$$[E_0] = [ES]\left[1 + \frac{K_s}{[S]} + \frac{k_2}{k_3}\right] \tag{5-8}$$

Rearranging gives

$$[ES] = \frac{[E]_0}{1 + K_s/S + k_2/k_3} \tag{5-9}$$

Since by assumptions (2) and (4) we know

$$v = \frac{d[P_1]}{dt} = \frac{d[P_2]}{dt} = k_2[ES] \tag{5-10}$$

we can derive Eq. 5-11 for the velocity of the observed reaction:

$$v = k_2\left[\frac{[E]_0}{1 + K_s/[S] + k_2/k_3}\right] = \frac{k_2[E]_0[S]}{K_s + [S](1 + k_2/k_3)} \tag{5-11}$$

Rearranging gives Eq. 5-11a:

$$v = \frac{[k_2/(1 + k_2/k_3)][E]_0[S]}{K_s/(1 + k_2/k_3) + [S]}$$

$$v = \frac{[k_2k_3/(k_2 + k_3)][E]_0[S]}{[K_sk_3/(k_2 + k_3)] + [S]} \tag{5-11a}$$

Comparison of Eqs. 5-11a and 5-1 shows that they are of the same kinetic form with

$$k_{cat} = \frac{k_2k_3}{k_2 + k_3} \quad \text{and} \quad K_{m(app)} = K_s\left(\frac{k_3}{k_2 + k_3}\right) \tag{5-12}$$

Thus one can see that the observation of Michaelis-Menten kinetics for an enzymic reaction does not mean that Eq. 5-2 uniquely holds and one must be cognizant of the possibility that Eq. 5-3 will apply. It is often naively assumed that K_m values give an indication of the tightness of the ES complex and that k_{cat} values give an indication of the organic catalytic power of the enzyme molecule. The fallacy of such assumptions has been discussed at some length in Chapter 4. Both of these indicators must be interpreted with caution. For example, it is seen in Eq. 5-12 that both k_{cat} and $K_{m(app)}$ may be complex quantities in the acyl-enzyme case and that changes in these quantities with variations in substrate structure may not be uniquely attributable to changes in rate of bond breaking or to changes in tightness of enzyme substrate association. It is only when the conditions can be so arranged that k_3 is uniquely determined in a system where k_{cat} and K_m are known that k_2 and K_s can be calculated. To achieve this generally requires the presteady-state treatment (see below) or the design of a special type of substrate. An example of such a substrate for chymotrypsin is p-nitrophenyl cinnamate where, according to the scheme of Eq. 5-3, P_1 is p-nitrophenol and P_2 is cinnamic acid. In independent experiments one could follow the rate of appearance of both P_1 and P_2 as long as deacylation (k_3) is rate determining.

For normal substrates, the general approach is unchanged: measure the rates of hydrolysis as a function of substrate concentration, keeping $[S]_0 \gg [E]_0$, and extract the rate constant by the appropriate (computerized) graphical method (e.g., Lineweaver-Burk or Eadie-Hofstee plot).

A commonly used measure of an enzyme's relative substrate specificity is the ratio of k_{cat}/K_m. From Eq. 5-12 it may be seen that $k_{cat}/K_{m(app)}$ is the

$$\frac{k_{cat}}{K_{m(app)}} = \frac{k_2}{K_s} \tag{5-13}$$

ratio of a rate constant and a dissociation constant for the particular enzymic process and thus comparisons of k_{cat}/K_m truly give a fair measure of an enzyme's overall preference for a particular substrate.

b. The Acyl-Enzyme Expression (Presteady-State)

The foregoing derivation of the kinetic expressions for Eq. 5-3 treated a system in which the steady-state had been reached. Generally, unless rapid reaction techniques (such as stopped-flow methods) are applied, only systems in the steady-state can be conveniently measured. This means that one is able to extract only the complex macroscopic rate constants k_{cat} and $K_{m(app)}$ from the rate data, and, as noted, this limits one's ability to interpret the data. As mentioned in Chapter 3, one may be able to determine the molar concentration of active sites with just steady-state kinetic observations on the appropriate substrate, but without the presteady-state treatment, the micro-scopic rate constants K_s, k_2, and k_3 are unattainable, and yet these are the important numbers that define the individual steps of the enzymatic reaction. The kinetic treatment presented next is general within the constraints of the assumptions for the steady-state treatment of the acyl-enzyme case and it will be seen that this treatment applies to the entire time course of an enzyme reaction following Eq. 5-3.

It will be noticed in Eq. 5-1 that if $K_m \gg [S]_0$, then pseudo-first-order kinetics will be followed, whereas when $[S]_0 \gg K_m$, pseudo-zero-order kinetics will obtain; thus the order of an enzyme reaction appears to vary if a wide enough range of substrate concentration is used. It was observed, however that during the hydrolysis of certain substrates (notably p-nitro-phenyl esters) by the enzyme chymotrypsin, biphasic behavior obtained in individual reactions (Figure 5-1). This biphasic appearance of p-nitrophenol was not explicable on the basis of the simple steady-state treatment and thus the following treatment was derived.

For this derivation we again assume that $[S]_0 \gg [E]_0$, that $d[ES']/dt \ll d[S]/dt$, and that $k_2(ES) = k_3(ES')$, and we retain Eqs. 5-4 and 5-5. From Eq. 5.3,

$$\mathrm{E + S} \underset{}{\overset{K_s}{\rightleftharpoons}} \mathrm{ES} \xrightarrow{k_2} \underset{\substack{+\\ \mathrm{P_1}}}{\mathrm{ES'}} \xrightarrow{k_3} \mathrm{E + P_2} \tag{5-3}$$

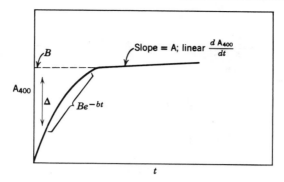

FIGURE 5-1. Appearance of pNP during hydrolysis of pNP esters by CT.

we know that

$$\frac{d[P_1]}{dt} = k_2[ES] \quad \text{and} \quad \frac{d[P_2]}{dt} = k_3[ES']$$

and that

$$\frac{d[ES']}{dt} = k_2[ES] - k_3[ES'] \tag{5-14}$$

In the early phases of the reaction $[S] \sim [S]_0$, and so using this and Eqs. 5-4 and 5-5 we may write

$$[ES] = \frac{[E]_0 - [ES']}{1 + K_s/[S]_0} \tag{5-15}$$

Our approach will be to obtain a rate expression for P_1 in terms of measurable parameters. As stated previously,

$$\frac{d[P_1]}{dt} = k_2[ES] \tag{5-16}$$

In Eq. 5-15 [ES'] is the only unknown quantity. If we now use Eq. 5-15 in 5-14, we get

$$\frac{d[ES']}{dt} = \frac{k_2[E]_0}{1 + K_s/[S]_0} - \frac{k_2[ES']}{1 + K_s/[S]_0} - k_3[ES']$$

$$\frac{d[ES']}{dt} = \frac{k_2[E]_0}{1 + K_s/[S]_0} - \left[k_3 + \frac{k_2}{1 + K_s/[S]_0}\right][ES'] \tag{5-17}$$

If we let

$$a = \frac{k_2[E]_0}{1 + K_s/[S]_0} \quad \text{and} \quad b = \left[k_3 + \frac{k_2}{1 + K_s/[S]_0} \right]$$

then we may write

$$\frac{d[ES']}{dt} = a - b[ES'] \tag{5-18}$$

This equation may be integrated to give

$$-\frac{1}{b} \ln (a - b[ES']) = t + c$$

which can be written as

$$bt = \ln \left(\frac{a}{a - b[ES']} \right) \quad \text{or} \quad e^{bt} = \frac{a}{a - b[ES']}$$

Therefore

$$[ES'] = \frac{a}{b} (1 - e^{-bt}) \tag{5-19}$$

With the aid of Eq. 5-15 it may be shown that at large values of t, Eq. 5-19 is equivalent to Eq. 5-7.

We now have in Eq. 5-19 an expression for [ES'] which contains no variables. This may be substituted into Eq. 5-15 to give

$$[ES] = \frac{[E]_0 - (a/b)(1 - e^{-bt})}{1 + K_s/[S]_0} \tag{5-20}$$

Therefore

$$\frac{d[P_1]}{dt} = \frac{k_2[E]_0 - k_2(a/b)(1 - e^{-bt})}{1 + K_s/[S]_0} \tag{5-21}$$

We now have in Eq. 5-21 an expression for [P$_1$] in terms of constant parameters a, b, k_2, and K_s and controllable parameters [E]$_0$ and [S]$_0$, and this may be integrated to give an expression that allows us to determine how the concentration of P$_1$ varies with time:

$$[P_1] = \frac{k_2[E]_0 - k_2(a/b)}{1 + K_s/[S]_0} t + \frac{(k_2)(a)}{b^2(1 + K_s/[S]_0)} (1 - e^{-bt}) \tag{5-22}$$

or more simply

$$[P_1] = At + B(1 - e^{-bt}) \tag{5-22a}$$

where

$$A = \frac{k_2[E]_0 - k_2(a/b)}{1 + K_s/[S]_0}$$

$$B = \frac{k_2(a)}{b^2(1 + K_s/[S]_0)}$$

Thus when t is very large, $[P_1] = At + B$, and this defines the linear portion (steady-state) of Figure 5-1, with slope $= A$ and intercept $= B$. Making the appropriate substitutions for a and b in the definitions of A and B, we find

$$A = \frac{[k_2k_3/(k_2 + k_3)][E]_0[S]_0}{[S]_0 + K_s[k_3/(k_2 + k_3)]} = \frac{k_{cat}[E]_0[S]_0}{[S]_0 + K_{m(app)}} \tag{5-23a}$$

$$B = [E]_0 \frac{[k_2/(k_2 + k_3)]^2}{(1 + K_{m(app)}/[S]_0)^2} \tag{5-23b}$$

Equation 5-23a is very similar to the Michaelis-Menten equation and may be treated similarly. Thus a plot of $1/A$ versus $1/[S]_0$ will be defined by

$$\frac{1}{A} = \frac{1}{k_{cat}[E]_0} + \frac{K_{m(app)}}{k_{cat}[E]_0[S]_0} \tag{5-24}$$

This will allow the calculation of k_{cat} and $K_{m(app)}$. Also, in the case where $k_3 \ll k_2$ and $K_{m(app)} \ll [S]_0$, B becomes equal to $[E]_0$, the active enzyme concentration. Thus from this linear portion of the curve we are able to determine the macroscopic constants k_{cat} and $K_{m(app)}$, and in favorable cases, we are able to calculate the molar concentration of the enzyme (from B) (see Figure 5-1).

When t is small (before the attainment of the steady state), $[P_1] = At + B - Be^{-bt}$. If one measures the distance between the experimental curve for these early times and the extrapolated portion of the linear region (differences Δ, Figure 5-1), one gets a series of Δ values defined by the expression $([P_1]_{extrap} - [P_1]_{early}) = Be^{-bt} = \Delta_t$, which defines a simple first-order relationship whose slope when plotted in the first-order fashion (Chapter 4) is given by $-b$. From our earlier definition of b we can rearrange to get

$$b = \frac{(k_2 + k_3)[S]_0 + k_3K_s}{K_s + [S]_0} \tag{5-25}$$

As written, Eq. 5-25 is difficult to treat, but it we note that

$$K_{m(app)} = (K_s) \frac{k_3}{k_2 + k_3} \tag{5-26}$$

and then stipulate that

$$[S]_0 \gg K_{m(app)} \tag{5-27}$$

then this means that

$$[S]_0 \gg \frac{k_3 K_s}{k_2 + k_3} \tag{5-28a}$$

or

$$k_3 K_s \ll (k_2 + k_3)[S]_0 \tag{5-28b}$$

Therefore

$$b = \frac{(k_2 + k_3)[S]_0}{[S]_0 + K_s} \tag{5-29}$$

$$\frac{1}{b} = \frac{1}{k_2 + k_3} + \frac{K_s}{k_2 + k_3} \cdot \frac{1}{[S]_0} \tag{5-30}$$

Again, we have a reciprocal relation that is linear in $1/[S]_0$ and thus from the determination of b as described above at a series of $[S]_0$ values one is able to calculate from the slope of a plot of $1/b$ versus $1/[S]_0$ the values of the term $K_s/(k_2 + k_3)$. But note now that $K_{m(app)}$ determined from the steady-state portion of the reaction is defined as in Eq. 5-12:

$$k_{cat} = \frac{k_2 k_3}{k_2 + k_3}; \qquad K_m = K_s \frac{k_3}{k_2 + k_3} \tag{5-12}$$

This expression for K_m differs from the expression for the slope of the $1/b$ versus $1/[S]_0$ plot by the factor k_3, and thus k_3 may readily be determined. With k_3 in hand, k_2 and K_s are easily obtained.

Thus by treating both the steady-state and the presteady-state of these biphasic reactions, K_s, k_2, and k_3 of Eq. 5-3 are all uniquely determined. It has often been the case that the presteady-state time scale is of the order of 5 seconds or less and this requires the use of rapid reaction techniques such as the stopped-flow method. It has turned out, however, for the case of p-nitrophenyl trimethylacetate plus α-chymotrypsin[4] that the presteady-state time scale is about 300 seconds, and thus this system can be conveniently handled on conventional spectrophotometers. The rate constants derived for that system are shown in Table 5-1.

The kinetic treatment on these pages provides very strong indirect evidence for the existence of the acyl enzyme. Direct evidence has come from the actual isolation of crystalline acyl enzymes (see Chapter 9).

TABLE 5-1 Kinetic Constants
for the System p-Nitrophenyl
Trimethylacetate + Chymotrypsin

$k_2 = 0.37\,(\pm 0.11)\,\text{sec}^{-1}$
$k_3 = 1.3\,(\pm 0.03) \times 10^{-4}\,\text{sec}^{-1}$
$K_s = 1.6\,(\pm 0.05) \times 10^{-3}M$
$k_{\text{cat}} = 1.3 \times 10^{-4}\,\text{sec}^{-1}$
$K_m = 5.6 \times 10^{-7}M$

pH Effects

Chemical modification studies, as discussed in Chapter 3, provide perhaps the most definitive identification of the groups on the enzyme responsible for its catalytic action; unfortunately they may not be routinely applicable. For example, after a multiple-step isolation of an enzyme, one may have only a few milligrams of active material, which is hardly sufficient for a thorough chemical modification study.

On the other hand, as long as one is reasonably certain that the enzyme preparation is homogeneous and that any impurities present do not interact with the enzyme, one can study its kinetic behavior as a function of pH and thereby get an excellent indication of the functional groups involved in the catalysis. This is only true, of course, for those catalytic groups that ionize, since, in theory, those groups that do not ionize in the pH range accessible to most enzymes (pH 2 to 10) are functionally independent of pH. Basically, one determines from a pH dependence study the pK_a values of those groups that "control" an enzyme's activity—that is, those groups whose association or dissociation causes the enzyme to manifest or not to manifest activity.

It should be noted that the ionizations detected in such a pH-rate study may be for groups that control an enzyme's conformation as well as for groups that participate in catalysis. This has been shown to be the case for the enzyme chymotrypsin, which has been found to have a bell-shaped dependence of rate as a function of pH. The acidic dissociation ($pK_a \sim 6.7$) is due to imidazole ring of histidine 57, and this group is involved in catalysis (see Chapter 9) in the dissociated form. The basic ionization (also a dissociation) has a pK_a of about 9 and is caused by the deprotonation of the α-amino group of isoleucine 16. In the protonated form this group has no catalytic function—it apparently controls the active conformation of the enzyme by participating in an ion-pair bond with the side chain carboxyl group of aspartic acid 194. When the α-amino group of Ile 16 dissociates, this ion-pair is broken and the side chain carboxylate of Asp 194 blocks the active site, thereby causing the enzyme to lose activity.

A fairly straightforward scheme useful in explaining the pH dependence of the kinetic behavior of a number of enzymes is that of Alberty and Massey:[5]

$$
\begin{array}{ccc}
EH_1H_2 & \xrightleftharpoons{K_{e1}} & EH_1 & \xrightleftharpoons{K_{e2}} & E \\
& & \updownarrow {\scriptstyle k_{-1} \| k_1} & & \\
EH_1H_2S & \xrightleftharpoons{K_{es1}} & EH_1S & \xrightleftharpoons{K_{es2}} & ES \\
& & \downarrow {\scriptstyle k_{cat}} & & \\
& & EH_1 + \text{Products} & &
\end{array}
\tag{5-31}
$$

In this scheme, only EH_1 can combine with the substrates S and only EH_1S can break down to products. Both the free enzyme and the enzyme-substrate complex have two distinct and important ionization constants controlling activity. Thus in Eq. 5-31, K_{e1} (K_{es1}) refers to a group on the free enzyme (enzyme-substrate complex) that must be dissociated for the enzyme to be active, while $K_{e2}(K_{es2})$ refers to a group on the enzyme (ES complex) that must be associated for the enzyme to be active. Here K_{e1} and K_{es1} refer to the ionizations of one prototropic group on the enzyme whose dissociation constant value shifts due to complex formation. The same is true for K_{e2} and K_{es2}. Generally, such complexation causes pK_a changes of a few tenths of a pH unit or less.

In the derivation of the equations that result from this approach, it is useful to have at hand the Michaelis pH functions. These functions are defined as follows:[6]

$$
f = 1 + \frac{K_1}{H} + \frac{K_1 K_2}{H^2}
\tag{5-32a}
$$

$$
f^- = 1 + \frac{H}{K_1} + \frac{K_2}{H}
\tag{5-32b}
$$

$$
f^= = 1 + \frac{H}{K_2} + \frac{H^2}{K_1 K_2}
\tag{5-32c}
$$

where $H \equiv [H^+]$. These equations apply to a substance which undergoes two successive ionizations, as in Eq. 5-33:

$$
AH_2 \xrightleftharpoons{K_1} AH_1{}^- \xrightleftharpoons{K_2} A^=
\tag{5-33}
$$

The reciprocals of these Michaelis pH functions are proportional to the amounts of the three forms of the substance (as in Eq. 5-33) present in solution at a given pH. Plots of these reciprocals versus pH are sigmoidal for Eqs. 5-32a and 5-32c and "bell-shaped" for Eq. 5-32b.

As shown in the scheme of Eq. 5-31 both EH_1 and EH_1S could be considered specific examples of the generalized substance AH_1^- in Eq. 5-33. One may consider $1/f^-$ to represent that fraction of the enzyme present in its catalytically active form (EH_1, EH_1S) with ionization in either direction to EH_1H_2 (EH_1H_2S) or $E(ES)$ giving an inactive form. This predicts bell-shaped pH-rate profiles for enzyme systems adhering to this scheme.

The maximum of the bell-shaped curve (pH_{opt}) appears at a pH halfway between pK_1 and pK_2, which means that

$$H_{opt} = \sqrt{K_1 K_2} \tag{5-34}$$

From Eqs. 5-32b and 5-34 it is easily shown that

$$\frac{1}{f_{opt}^-} = \frac{1}{1 + 2\sqrt{K_2/K_1}} \tag{5-35}$$

The value of $[H^+]$ at which $(1/f^-) = \frac{1}{2}(1/f_{opt}^-)$ is given by

$$\frac{1}{f^-} = \frac{1}{2 + 4\sqrt{K_2/K_1}} = \frac{K_1 H}{K_1 H + H^2 + K_1 K_2} \tag{5-36}$$

This simplifies to the quadratic expression

$$H^2 - (K_1 + 4H_{opt})H + H_{opt}^2 = 0 \tag{5-37}$$

if we recall that $H_{opt} = \sqrt{K_1 K_2}$. This equation has two real roots, H_1 and H_2 corresponding to the half-maximal points on the two sides of the optimum. Since in the quadratic formula the roots are given by $x_1, x_2 = (-b \pm \sqrt{b^2 - 4ac})/2a$, $(ax^2 + bx + c = 0)$, it may be seen that

$$H_1 + H_2 = K_1 + 4H_{opt} \tag{5-38}$$

This allows the determination of K_1 and Eq. 5-34 then allows the determination of K_2. This is a general treatment giving values of dissociation constants which refer to the particular kinetic parameter whose behavior as a function of pH is being studied. One may easily obtain the equations from which the theoretical pH-rate curves for enzyme reactions are drawn by assuming that, as shown in Eq. 5-31, only EH_1 binds substrate and only EH_1S breaks down to products, and the ionizations shown are much faster than complex formation or catalysis, so that equilibrium is maintained (horizontally in Eq. 5-31) between the various forms. Then, since at any pH the fraction of catalytically active form (EH_1; EH_1S) is given by $1/f_{eH_1}^-$ or $1/f_{eH_1S}^-$, one may write

$$V_{max} = k_{cat}[E]_0 \frac{1}{f_{es}^-} = \frac{(k_{cat})_{opt}[E]_0}{1 + H/K_{es1} + K_{es2}/H} \tag{5-39}$$

Thus Eqs. 5-40, 5-41, and 5-42a describe the theoretical pH dependence of the parameters k_{cat}, k_{cat}/K_m and K_m:

$$k_{cat} = \frac{(k_{cat})_{opt}}{1 + H/K_{es1} + K_{es2}/H} \tag{5-40}$$

$$k_{cat}/K_m = \frac{(k_{cat}/K_m)_{opt}}{1 + H/K_{e1} + K_{e2}/H} \tag{5-41}$$

$$K_m = (K_m)_{opt} \left[\frac{1 + H/K_{e1} + K_{e2}/H}{1 + H/K_{es1} + K_{es2}/H} \right] \tag{5-42a}$$

In these equations $(k_{cat})_{opt}$, $(K_m)_{opt}$, and $(k_{cat}/K_m)_{opt}$ are the values of these particular parameters at H_{opt}. Thus these equations merely mean that the value of k_{cat} (or K_m or k_{cat}/K_m) is equal to its theoretically optimum value multiplied by that fraction of the catalytically active species present at the particular pH value. Since V_{max} depends entirely on the concentration of enzyme-substrate complex, it is readily apparent (Eqs. 5-39 and 5-40) that the ionization constants derived from a study of the pH dependence of k_{cat} refer to ionizations on the complex. Similarly, K_m involves both free and complexed enzyme, as shown in Eq. 5-42b, which is actually the same as 5-42a.

$$K_m = \frac{[E][S]}{[ES]} = \frac{(1/f_{es}{}^-)[E]_0[S]}{(1/f_e{}^-)[ES]} = \frac{f_e{}^-}{f_{es}{}^-} (K_m)_{opt} \tag{5-42b}$$

Thus the ionization constants derived from a study of the pH dependence of K_m are not easily attributable to one or another form of the enzyme. However, since $k_{cat} = (k_{cat})_{opt}/f_{es}{}^-$, by substitution, we obtain

$$k_{cat}/K_m = \frac{(k_{cat})_{opt}/f_{es}{}^-}{(K_m)_{opt} f_e{}^-/f_{es}{}^-} = \frac{(k_{cat}/K_m)_{opt}}{f_e{}^-} \tag{5-42c}$$

Therefore the ionization constants extracted from a study of the pH dependence of k_{cat}/K_m refer to ionizations taking place on the free, uncomplexed enzyme. The treatment of Eq. 5-31 makes the final assumption that the *substrate* is either (1) undissociated or (2) 100% in one ionic form, or (3) all ionic forms have the same affinity for the enzyme. However, if the substrate can ionize, it is quite likely that assumption (3) will not hold, that is, the enzyme will show different affinities for the different ionic states of the substrate. This requires that the theoretical expression for K be corrected by a factor to account for this substrate ionization. This factor is obtained through the evaluation of the appropriate Michaelis pH function for the substrate,

$$f_s{}^- = 1 + \frac{K_{1S}}{H} + \frac{H}{K_{2S}} \tag{5-43}$$

which is given by Eq. 5-43. Here K_{1S} and K_{2S} are the ionization constants for the process

$$H_1H_2S \overset{K_{1s}}{\rightleftharpoons} H_1S \overset{K_{2s}}{\rightleftharpoons} S$$

Incorporation of this factor changes Eqs. 5-41 and 5-42 to Eqs. 5-44 and 5-45, respectively. Equation 5-40 is unaffected.

$$k_{cat}/K_m = \frac{(k_{cat}/K_m)_{opt}}{[1 + H/K_{e1} + K_{es}/H][1 + K_{1s}/H + H/K_{2s}]} \tag{5-44}$$

$$K_m = (K_m)_{opt} \left[\frac{1 + H/K_{e1} + K_{e2}/H}{1 + H/K_{es1} + K_{es2}/H} \right] \left[1 + \frac{K_{1s}}{H} + \frac{H}{K_{2s}} \right] \tag{5-45}$$

The use of f_s^- in Eq. 5-43 implies that the H_1S form of the substrate combines with the enzyme. If some other form of the substrate combines, then the appropriate Michaelis function must be used.*

Experimentally, one usually determines k_{cat} and K_m at a variety of pH values and then plots these values, or their logarithms, versus pH in order to determine the appropriate pK_a values. As long as the species involved are properly represented by the Michaelis pH functions, the curves will appear

---

* The scheme of Eq. 5-31 involves only three forms of the enzyme. Although Eq. 5-31 has proven quite useful in practice, a more realistic scheme would be that of 5-i.

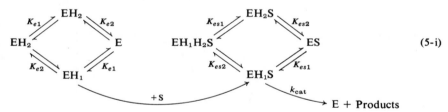

$$(5\text{-}i)$$

This scheme allows the ionizations of H_1 and H_2 to be independent of each other rather than requiring EH_1H_2 to ionize to EH_1 before ionizing to E. This scheme does not go so far as to recognize that the ionization of one group will affect the pK_a of the other group; the dissociation constant for H_2 ionization is K_{e2} whether the ionizing species is EH_2 or EH_1H_2. The net effect of this change is to add a constant term K_2/K_1 to the pH functions governing the amounts of the various species in solution and lead to Eq. 5-ii and 5-iv describing the pH dependence of the enzymatic reaction.

$$k_{cat} = \frac{(k_{cat})_{opt}}{1 + H/K_{es1} + K_{es2}/H + K_{es2}/K_{es1}} \tag{5-ii}$$

$$k_{cat}/K_m = \frac{(k_{cat}/K_m)_{opt}}{1 + H/K_{e1} + K_{e2}/H + K_{e2}/K_{e1}} \tag{5-iii}$$

$$K_m = (K_m)_{opt} \left[\frac{1 + H/K_{e1} + K_{e2}/H + K_{e2}/K_{e1}}{1 + H/K_{es1} + K_{es2}/H + K_{es2}/K_{es1}} \right] \tag{5-iv}$$

as in Figure 5-2 (or Figure 5-3 for logarithm values—using f^- as the example). In Figure 5-3 the three phases of the curve correspond to the pH region where one of the three terms of the pH function is dominant. This will be obvious if one computes the values of f^- at pH 3, 7, and 12, taking $K_1 = 10^{-5}M$ and $K_2 = 10^{-10}M$. Moreover, the slopes of the various phases equal the

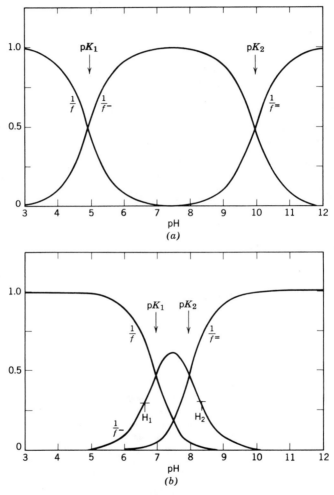

FIGURE 5-2.   pH Dependence of the pH functions. (a) Plot of the reciprocals of the pH functions versus pH for a system with two ionizing groups with pK_a values of 5 and 10. (b) Plot of the reciprocals of the pH functions versus pH for a symmetrical system with two ionizing groups with pK_a values of 7 and 8. (After M. Dixon and E. C. Webb, *Enzymes*, 2nd ed., Academic, New York, 1964.)

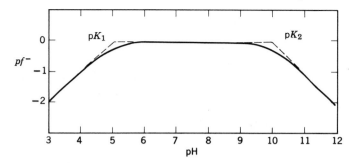

FIGURE 5-3. Plot of $pf^- = -\log f^- = \log 1/f^-$ for the system of Figure 5-2a. (After M. Dixon and E. C. Webb, *Enzymes*, 2nd ed., Academic, New York, 1964.)

power of hydrogen occurring in the term dominant in that phase (i.e., 1, 0, or -1 for the three terms of f^-).

As pointed out earlier, it is common practice in enzyme studies to attribute the values of K_1 and K_2 determined by experiment to specific groups on the enzyme, with K_{e1} and K_{e2} referring to the free enzyme and K_{es1} and K_{es2} referring to the enzyme-substrate complex (Eq. 5-31). Although this may indeed be the case, strictly speaking these K values refer to "stages" of ionization of the species involved. This distinction must be made because there are many acidic and basic groups on an enzyme and their ionizations are somewhat interdependent.

One approach used to substantiate a conclusion as to the identity of an ionizing group based on its kinetically determined pK_a value is to determine the $\Delta H°$ for the ionization ($\Delta H_i°$). This is done by studying the pH-rate dependence at a series of different pK_a values as a function of temperature. One will recall from thermodynamic principles that

$$\Delta H_i° = (2.303RT^2)\frac{d\log(k_{+1}/k_{-1})}{dT} \tag{5-46}$$

$$pK_a = -\log K_a = -\log\frac{k_{-1}}{k_{+1}} = \log\frac{k_{+1}}{k_{-1}} \tag{5-47}$$

for the reaction

$$HA \underset{k_{-1}}{\overset{k_{+1}}{\rightleftharpoons}} H^\oplus + A^\ominus$$

It may be seen therefore, that

$$\Delta H_i° = (2.303RT^2)\frac{d(pK_a)}{dT}$$

or

$$\left(\frac{\Delta H_i{}^\circ}{2.303RT^2}\right) dt = d(\mathrm{p}K_a) \qquad (5\text{-}48)$$

Thus a plot of $\mathrm{p}K_a$ versus $1/T$ should give a straight line of slope $\Delta H_i{}^\circ/2.303R$. The values of $\Delta H_i{}^\circ$ so obtained may be compared with the values in Table 5-2. The combined knowledge of $\mathrm{p}K_a$ and $\Delta H_i{}^\circ$ should allow a reasonably firm identification of the ionizing group.

TABLE 5-2 pK Values and Heats of Ionization of Some Groups Present in Proteins[a]

Group	pK_a (25°C)	ΔH_i (kcal/mole)
Carboxyl (α)	3.0–3.2	±1.5
Carboxyl (aspartyl, glutamyl)	3.0–4.7	±1.5
Imidazolium (histidine)	5.6–7.2	6.9–7.5
Sulfhydryl	8.3–8.6	—
Phenolic hydroxyl (Tyrosine)	9.8–10.4	6.0
Ammonium (α)	7.6–8.4	10–13
Ammonium (α, cystine)	6.5–8.5	—
Ammonium (ε, lysine)	9.4–10.6	10–12
Guanidinium (arginine)	11.6–12.6	12–13

[a] Source: M. Dixon and E. C. Webb, *Enzymes*, 2nd ed., Academic Press, New York, 1964, Chap. IV. p. 144.

References

1. B. Zerner and M. L. Bender, *J. Am. Chem. Soc.*, **86**, 3669 (1964), and accompanying papers.
2. E. Zeffren and E. T. Kaiser, *Arch. Biochem. Biophys.*, **126**, 965 (1968).
3. E. Zeffren and E. T. Kaiser, *J. Am. Chem. Soc.*, **89**, 4204 (1967).
4. M. L. Bender, F. J. Kezdy, and F. C. Wedler, *J. Chem. Educ.*, **44**, 84 (1967).
5. R. A. Alberty and V. Massey, *Biochem. Biophys. Acta*, **13**, 347 (1954).
6. M. Dixon and E. C. Webb, *Enzymes*, 2nd ed., Academic, New York, 1964, Chap. IV.

CHAPTER 6

Kinetics III—Reversible Inhibition of Enzyme Action

Enzyme Specificity[1]

Before we discuss the various types of inhibition, the subject of specificity will be briefly mentioned, since the same factors that dictate which compounds serve as substrates for an enzyme also dictate which compounds serve as inhibitors. This is clearly seen in those cases where the inhibitor is structurally analogous to the substrate. For example, the dipeptide N-acetyl-L-phenyl-alanyl-L-phenylalanine is a substrate of the enzyme pepsin and has an apparent Michaelis constant of about $10^{-3}M$, whereas the analogous compound N-acetyl-D-phenylalanyl-D-phenylalanine is a competitive inhibitor with a dissociation constant for the E-I complex of about $10^{-3}M$.

Thus the study of an enzyme's substrate and inhibitor specificities is the study of the structural parameters that determine whether a molecule will form a complex with the enzyme. It is obvious from these considerations that specificity will be dictated not only by the thermodynamics of the molecular interactions involved but also by the native structure of the enzyme. The Michaelis and inhibition constants above are energetically equivalent and yet one molecule is a substrate for pepsin and the other is an inhibitor.

The example of enzyme specificity just mentioned may be termed *configurational* or *steric specificity*, that is, enzymes possess the capacity to distinguish between enantiomers and diastereomers. In some cases, as above, the stereo-isomers that are not substrates are found to be inhibitors; in other cases this is not so.

An even more subtle manifestation of specificity is the selection of one of two apparently identical moieties bound to a "symmetrical" carbon atom. This suggestion is originally due to Ogston,[2] who noted that for a molecule of structure CXX'YZ, which possesses a plane of symmetry defined by the Y-C-Z atoms, an enzyme could distinguish between the two X atoms if it had specific binding sites for three of the four groups bound to the central carbon atom. This is schematically shown in Figure 6-1. With a distinct

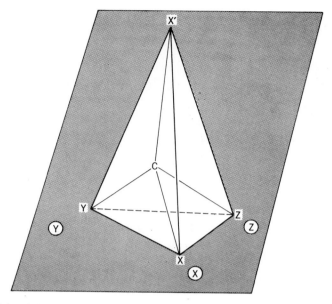

FIGURE 6-1. Schematic representation of the stereospecific three-point attachment of species CXX'YZ to an enzyme.

binding site for groups X, Y, and Z, the location of group X' is determined uniquely. Since the relationship between binding sites $\widehat{x}$, $\widehat{y}$, and $\widehat{z}$ is fixed in the enzyme's active structure, the placement of X' at site $\widehat{x}$ will necessarily displace Y and Z from their sites leading to either no binding or unproductive binding. This type of specificity has been shown to occur in enzymic systems and the area has been recently reviewed by Popjak,[3] who has demonstrated that even a three-point attachment hypothesis is overrestrictive if a symmetrical substrate can have dissymmetry imposed upon it by steric factors in the binding process. This of course still requires a CXX'YZ type of substrate.

Another important type of specificity is relative reaction rate or kinetic specificity. That is, for a series of substrates, some will react more rapidly than others. In this case, the enzyme is said to be more specific for structures resembling the rapidly reacting species. It was pointed out in Chapter 5 that the readily obtainable kinetic constant k_{cat}/K_m is equal to the ratio k_2/K_s, a simple rate constant and a simple dissociation constant. Thus this number can be considered as a reliable measure of an enzyme's kinetic preference for a given substrate—the larger k_{cat}/K_m, the more preferred is the substrate. A detailed discussion of the use of this specificity parameter has been given for the enzyme chymotrypsin, and the interested reader is referred to Bender and Kezdy[4] (see also Chapter 9).

As the reader may have surmised from this brief discussion, the chemical sophistication of an enzyme's specificity is one of its most outstanding characteristics. The detailed chemistry of specificity is still poorly understood.

Types of Reversible Inhibition

The equations that describe the most common kinds of reversible inhibition —competitive inhibition (full and partial), noncompetitive inhibition (full and partial), and uncompetitive inhibition—will be derived in this section. The meaning behind these terms will become apparent as they are discussed. Reversible inhibition implies by its name that the inhibiting species can be removed from the enzyme, restoring full activity to it. The means of removal of the inhibitor are immaterial; dialysis, chromatographic methods, electrophoresis, and other methods may be used as long as the process does not harm the native enzyme. Irreversible inhibitors are more appropriately called inactivators and have been discussed in Chapter 3.

a. Fully Competitive Inhibition

In Chapters 4 and 5 it was seen that no matter what detailed scheme was treated, the gross rate expression for the macroscopic constants k_{cat} and K_m was the same. Thus the classical Michaelis-Menten treatment, the steady-state treatment of Briggs and Haldane, or the acyl-enzyme treatment all led to a rate expression of the form

$$v = \frac{k_{cat}[E]_0[S]}{K_m + [S]} \tag{6-1}$$

This is most simply derived from

$$E + S \underset{k_{-1}}{\overset{k_1}{\rightleftharpoons}} ES \overset{k_2}{\longrightarrow} E + P \tag{6-2}$$

Consider an inhibitor molecule I which can also combine with E in such a way that EI can no longer bind S:

$$E + I \underset{k_{-i}}{\overset{k_i}{\rightleftharpoons}} EI \tag{6-3}$$

If we further stipulate that ES cannot bind I, then Eqs. 6-2 and 6-3 describe the conditions of fully competitive inhibition: E will bind only I or only S, but not both; only ES will break down to products. In this case $[E]_0$ is given by

$$[E]_0 = [E] + [ES] + [EI] \tag{6-4}$$

and K_i is defined by

$$K_i = \frac{[E][I]}{[EI]} = \frac{k_{-i}}{k_i}. \tag{6-5}$$

Thus

$$[EI] = \frac{[E][I]}{K_i} \tag{6-6a}$$

$$[ES] = \frac{[E][S]}{K_m} \tag{6-6b}$$

Substituting these in Eq. 6-4 gives

$$[E]_0 = [E]\left(1 + \frac{[S]}{K_m} + \frac{[I]}{K_i}\right)$$

$$[E] = \frac{[E]_0}{1 + [S]/K_m + [I]/K_i} \tag{6-7}$$

Therefore in Eq. 6-6b, we can substitute for [E] and get

$$(ES) = \frac{[E]_0[S]}{K_m(1 + [S]/K_m + [I]/K_i)} = \frac{[E]_0[S]}{[S] + K_m(1 + [I]/K_i)} \tag{6-8}$$

Since we recall from Chapter 4 that $v = k_2[ES]$, we get

$$v = \frac{k_2[E_0][S]}{K_m(1 + [I]/K_i) + [S]} \tag{6-9}$$

Equation 6-9 is the expression that describes fully competitive inhibition.* It is observed that this expression is of the same form as Eq. 6-1 except that now the apparent K_m has a different meaning. Thus, if one knows K_m from a series of experiments performed in the absence of a fully competitive inhibitor, then when the experiments are repeated in the presence of the inhibitor Eq. 6-9 says that only K_m will change, and it will change by the factor $(1 + [I]/K_i)$, and V_{max} will be unaffected. Before one can use this approach to the determination of K_i, however, one must determine if the inhibition is indeed competitive. How this is done will be explained later in this chapter.

b. Partially Competitive Inhibition

In this type of inhibition, E may combine with I or S as before but also EI may combine with S and ES may combine with I to give the same EIS

---

* Equation 6-9 holds no matter what the kinetic significance of K_m is.

complex. Both EIS and ES break down to products at the same rate. Thus we have the following equations:

$$E + S \underset{k_{-1}}{\overset{k_1}{\rightleftharpoons}} ES \overset{k_2}{\longrightarrow} E + P \qquad K_m = \frac{[E][S]}{[ES]} \qquad (6\text{-}2)$$

$$E + I \underset{k_{-i}}{\overset{k_i}{\rightleftharpoons}} EI \qquad\qquad K_i = \frac{[E][I]}{[EI]} \qquad (6\text{-}10a)$$

$$EI + S \underset{k'_{-i}}{\overset{k'_i}{\rightleftharpoons}} EIS \qquad\qquad K'_i = \frac{[EI][S]}{[EIS]} \qquad (6\text{-}10b)$$

$$ES + I \underset{k'_{-i}}{\overset{k'_i}{\rightleftharpoons}} EIS \qquad\qquad K''_i = \frac{[ES][I]}{[EIS]} \qquad (6\text{-}10c)$$

$$EIS \overset{k_2}{\longrightarrow} EI + P \qquad\qquad (6\text{-}10d)$$

$$[E]_0 = [E] + [ES] + [EI] + [EIS] \qquad (6\text{-}11)$$

The velocity of this reaction will be given by

$$v = k([ES] + [EIS]) \qquad (6\text{-}12)$$

From the definitions of the constants K_m, K_i, and K'_i we may write

$$[ES] = \frac{[E][S]}{K_m} \qquad (6\text{-}13a)$$

$$[EI] = \frac{[E][I]}{K_i} \qquad (6\text{-}13b)$$

$$[EIS] = \frac{[EI][S]}{K'_i} \qquad (6\text{-}13c)$$

Therefore we can get

$$[ES] + [EIS] = [E]\left(\frac{[S]}{K_m} + \frac{[I][S]}{K_i K'_i}\right) \qquad (6\text{-}14)$$

From Eqs. 6-13 and 6-11 we can derive

$$[E] = \frac{[E]_0}{1 + [I]/K_i + [S]/K_m + [I][S]/K_i K'_i} \qquad (6\text{-}15)$$

Substituting for [E] in Eq. 6-14, we get

$$[ES] + [EIS] = \frac{[E]_0[S](1/K_m + I/K_iK_i')}{1 + [I]/K_i + [S]/K_m + [I][S]/K_iK_i'} \qquad (6\text{-}16)$$

$$= \frac{[E]_0[S](1/K_m + [I]/K_iK_i')K_m}{K_m(1 + [I]/K_i) + [S](1 + [I]K_m/K_iK_i')}$$

We are thus led to Eq. 6-17 as the rate expression

$$v = \cfrac{k_2[E]_0[S]}{\cfrac{K_m(1 + [I]/K_i)}{K_m(1/K_m)(1 + [I]K_m/K_iK_i')} + \cfrac{[S](1 + [I]K_m/K_iK_i')}{1/K_m(1 + [I]K_m/K_iK_i')K_m}}$$

$$\qquad (6\text{-}17)$$

$$= \cfrac{k_2[E]_0[S]}{K_m\left[\cfrac{1 + [I]/K_i}{1 + [I]K_m/K_iK_i'}\right] + [S]}$$

It should be obvious that by varying the substrate concentration in a series of experiments at a constant inhibitor concentration one will be unable to distinguish this form of inhibition from the purely competitive type. To do this requires a series of experiments with a variable inhibitor concentration at a fixed substrate concentration. At very large values of [I], we get

$$v = \frac{k_2[E_0][S]}{K_i' + [S]} \qquad (6\text{-}18)$$

Thus if one knows one is dealing with a partial competitive inhibition and working at a level of [I] such that Eq. 6-18 holds, then K_i' may be determined. Then if K_m has been previously determined from experiments in the absence of inhibitor, K_i may be determined from experiments in the range of inhibitor concentrations where Eq. 6-17 holds. This is complex behavior and requires a fair amount of effort to elucidate fully.

c. Fully Noncompetitive Inhibition

In this type of inhibition the affinity of the enzyme for the substrate is not affected by complexation with the inhibitor, and vice versa. Moreover, the pure noncompetitive inhibition is characterized by having only ES, not EIS, break down to products. In the partially noncompetitive case EIS can break down to products and does so at a rate different from that of ES. The equations describing purely noncompetitive behavior are:

$$E + S \underset{k_{-1}}{\overset{k_1}{\rightleftharpoons}} ES \overset{k_2}{\longrightarrow} E + P \qquad K_m = \frac{[E][S]}{[ES]} = \frac{[EI][S]}{[EIS]} \qquad (6\text{-}19a)$$

$$E + I \underset{k_{-i}}{\overset{k_i}{\rightleftharpoons}} EI$$

$$K_i = \frac{[E][I]}{[EI]} = \frac{[ES][I]}{[EIS]} \tag{6-19b}$$

$$EI + S \overset{K_m}{\rightleftharpoons} EIS \tag{6-19c}$$

$$ES + I \overset{K_i}{\rightleftharpoons} EIS \tag{6-19d}$$

The conservation equation here is

$$[E]_0 = [E] + [ES] + [EI] + [EIS] \tag{6-20}$$

From Eq. 6-19 we may write

$$[ES] = \frac{[E][S]}{K_m} \tag{6-21a}$$

$$[EI] = \frac{[E][I]}{K_i} \tag{6-21b}$$

$$[EIS] = \frac{[ES][I]}{K_i} = \frac{[E][S][I]}{K_m K_i} \tag{6-21c}$$

Therefore we may solve for [E] and rewrite Eq. 6-20 as

$$[E] = \frac{[E]_0}{1 + [S]/K_m + [I]/K_i + [S][I]/K_m K_i} \tag{6-22}$$

Since we know by our initial restrictions that

$$v = k_2[ES] = k_2 \frac{[E][S]}{K_m} \tag{6-23}$$

we may then write

$$v = \frac{k_2[E]_0[S]}{K_m(1 + [S]/K_m + [I]/K_i + [S][I]/K_m K_i)} \tag{6-24}$$

$$= \frac{k_2[E]_0[S]}{K_m(1 + [I]/K_i) + [S](1 + [I]/K_i)}$$

Dividing numerator and denominator by $(1 + [I]/K_i)$ gives

$$v = \frac{\{k_2[E]_0/(1 + [I]/K_i)\}[S]}{K_m + [S]} \tag{6-25}$$

Thus it is seen that in purely noncompetitive inhibition V_{max} is changed whereas K_m is unaffected. This is just the opposite from what occurs in purely competitive inhibition. The Lineweaver-Burk plot of purely noncompetitive behavior shows a changing ordinate intercept for a series of reactions at fixed inhibitor concentrations and variable substrate concentration.

d. Partially Noncompetitive Inhibition

In this case Eqs. 6-19a to 6-19d still apply, but they must be supplemented by Eq. 6-19e:

$$ \text{EIS} \xrightarrow{k_2'} \text{EI} + \text{P} \tag{6-19e} $$

Thus the overall velocity of the enzymic reaction in this situation is

$$ v = k_2[\text{ES}] + k_2'[\text{EIS}] \tag{6-26} $$

Through a derivation analogous to that leading to Eq. 6-25, one can derive

$$ v = \frac{k_2[\text{E}]_0[\text{S}]/(1 + [\text{I}]/K_i)}{K_m + [\text{S}]} + \frac{k_2'[\text{E}]_0[\text{S}]([\text{I}]/K_i)/(1 + [\text{I}]/K_i)}{K_m + [\text{S}]} \tag{6-27} $$

which may be rewritten

$$ v = \frac{\dfrac{k_2[\text{E}]_0[\text{S}] + k_2'[\text{E}]_0[\text{S}][\text{I}]/K_i}{1 + [\text{I}]/K_i}}{K_m + [\text{S}]} \tag{6-28} $$

When [I] is very large, $V_{\max} = k_2'[\text{E}]_0$, allowing determination of k_2'. Determination of K_i is somewhat involved, although less so than for the partially competitive case. Here, K_i can be calculated from the ordinate intercept as long as k_2' has been determined (see the section on determining inhibitor constants in this chapter).

e. Uncompetitive Inhibition

This type of inhibition is characterized by complex formation of the inhibitor only with ES. The native enzyme E will not form a complex with the inhibitor. This type of inhibition is fairly rare in monomeric, single-substrate enzymes but occurs more frequently in more complex systems. The equations that describe this scheme are as follows:

$$ \text{E} + \text{S} \underset{k_{-1}}{\overset{k_1}{\rightleftharpoons}} \text{ES} \xrightarrow{k_2} \text{E} + \text{P} \qquad K_m = \frac{[\text{E}][\text{S}]}{[\text{ES}]} \tag{6-2} $$

$$ \text{ES} + \text{I} \underset{k_{-i}}{\overset{k_i}{\rightleftharpoons}} \text{EIS} \qquad K_i = \frac{[\text{ES}][\text{I}]}{[\text{EIS}]} \tag{6-29} $$

The conservation equation is

$$ [\text{E}]_0 = [\text{E}] + [\text{ES}] + [\text{EIS}] \tag{6-30} $$

The rate is given by

$$ v = k_2[\text{ES}] = \frac{k_2[\text{E}][\text{S}]}{K_m} \tag{6-31} $$

Using the definitions of K_m and K_i and Eq. 6-30 we may solve for [E]:

$$[E] = \frac{[E]_0}{1 + [S]/K_m + [S][I]/K_m K_i} \qquad (6\text{-}32)$$

Thus we get as the rate expression

$$v = \frac{k_2[E]_0[S]}{K_m + S(1 + [I]/K_i)} \qquad (6\text{-}33)$$

The double reciprocal Lineweaver-Burk plot of this treatment (see below) shows an unchanged slope and intercepts that get larger as [I] get larger. Thus if one knows V_{max} (i.e., k_2) from experiments in the absence of inhibitor then K_i is easily calculated. Dixon and Webb[1] have pointed out that for a noncompetitive inhibitor of an enzyme acting on a substrate for which $k_2 \gg k_{-1}$, the Lineweaver-Burk plots will be indistinguishable from those of uncompetitive inhibition, even though E can combine with I.

Determination of Inhibitor Constants

Perhaps the most widely used approach to the determination of inhibitor constants is the Lineweaver-Burk double reciprocal plot. From the slope and intercepts of these plots drawn for experiments run in the presence and absence of inhibitor, one often (but, as pointed out, not always) may determine k_{cat} (or V_{max}), K_m, and K_i. This can best be seen by presenting the reciprocal relations as they would be used, and this is done in Table 6-1.

Figure 6-2 represents Lineweaver-Burk plots for the types of inhibition exemplified in Table 6-1. It is apparent from the table and the figure that K_i is readily determined for cases a, c, and e. One merely studies the enzymic reaction at a series of substrate concentrations at a fixed inhibitor concentration and draws the appropriate double reciprocal plot. The algebra then allows a straightforward calculation of K_i. As pointed out earlier, however, more involved studies are necessary to determine K_i values for cases b and d.

Other methods exist for the determination of inhibitor constants, and the reader is referred to more comprehensive texts in enzyme kinetics for an explanation of them. One other method will be presented here, and that is the method of Dixon.[5] This is a simple graphical method that requires only the determination of v at two or three substrate concentrations using a series of inhibitor concentrations at each substrate level. A plot of $(1/v)$ versus [I] for each set of reactions will give a series of lines that will intersect at a value of [I] equal to $-K_i$; moreover, the noncompetitive inhibition intersection point will be on the abscissa while the competitive inhibition intersection point will be above it. This can be shown as follows.

TABLE 6-1 Lineweaver-Burk Relations and Intercept and Slope Definitions

Type	Equation	Ordinate Intercept	Slope	Abscissa Intercept
No inhibitor	$\dfrac{1}{v} = \dfrac{K_m}{k_2[E]_0}\left(\dfrac{1}{[S]}\right) + \dfrac{1}{k_2[E]_0}$	$\dfrac{1}{k_2[E]_0}$	$\dfrac{K_m}{k_2[E]_0}$	$\dfrac{-1}{K_m}$
a	$\dfrac{1}{v} = \dfrac{K_m\left(1+\frac{[I]}{K_i}\right)}{k_2[E]_0}\left(\dfrac{1}{[S]}\right) + \dfrac{1}{k_2[E]_0}$	$\dfrac{1}{k_2[E]_0}$	$\dfrac{K_m\left(1+\frac{[I]}{K_i}\right)}{k_2[E]_0}$	$-\dfrac{1}{K_m\left(1+\frac{[I]}{K_i}\right)}$
b	$\dfrac{1}{v} = \dfrac{K_m\left[\left(1+\frac{[I]}{K_i}\right)\Big/\left(1+\frac{[I](K_m)}{K_iK_i'}\right)\right]}{k_2[E]_0}\left(\dfrac{1}{[S]}\right) + \dfrac{1}{k_2[E]_0}$	$\dfrac{1}{k_2[E]_0}$	$\dfrac{K_m}{k_2[E]_0}\left[\dfrac{1+\frac{[I]}{K_i}}{1+\frac{[I](K_m)}{K_iK_i'}}\right]$	$-\dfrac{\left[1+\frac{[I](K_m)}{K_iK_i'}\right]}{K_m\left(1+\frac{[I]}{K_i}\right)}$
c	$\dfrac{1}{v} = \dfrac{K_m\left(1+\frac{[I]}{K_i}\right)}{k_2[E]_0}\left(\dfrac{1}{[S]}\right) + \dfrac{1+\frac{[I]}{K_i}}{k_2[E]_0}$	$\dfrac{1+\frac{[I]}{K_i}}{k_2[E]_0}$	$\dfrac{K_m\left(1+\frac{[I]}{K_i}\right)}{k_2[E]_0}$	$-\dfrac{1}{K_m}$
d	$\dfrac{1}{v} = \dfrac{K_m\left(1+\frac{[I]}{K_i}\right)}{k_2[E]_0 + \frac{k_2'[E]_0[I]}{K_i}}\left(\dfrac{1}{[S]}\right) + \dfrac{1+\frac{[I]}{K_i}}{k_2[E]_0 + \frac{k_2'[E]_0[I]}{K_i}}$	$\dfrac{1+\frac{[I]}{K_i}}{k_2[E]_0 + \frac{k_2'[E]_0[I]}{K_i}}$	$\dfrac{K_m\left(1+\frac{[I]}{K_i}\right)}{k_2[E]_0 + \frac{k_2'[E]_0[I]}{K_i}}$	$-\dfrac{1}{K_m}$
e	$\dfrac{1}{v} = \dfrac{K_m}{k_2[E]_0}\left(\dfrac{1}{[S]}\right) + \dfrac{1+\frac{[I]}{K_i}}{k_2[E]_0}$	$\dfrac{1+\frac{[I]}{K_i}}{k_2[E]_0}$	$\dfrac{K_m}{k_2[E]_0}$	$-\dfrac{1+\frac{[I]}{K_i}}{K_m}$

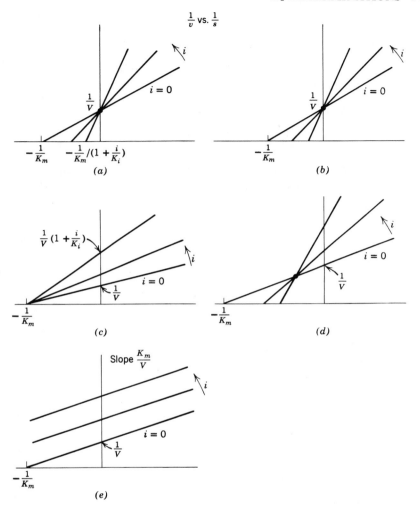

FIGURE 6-2.   Lineweaver-Burk plots of the various types of inhibition. (*a*) Fully competitive. (*b*) Partially competitive. (*c*) Fully noncompetitive. (*d*) Partially noncompetitive. (*e*) Uncompetitive.

For competitive inhibition we have

$$\frac{1}{v} = \frac{K_m}{V_{\max}} \frac{1}{[S]} + \frac{K_m}{V_{\max}} \frac{[I]}{K_i} \frac{1}{[S]} + \frac{1}{V_{\max}} \tag{6-34}$$

We want to plot $(1/v)$ versus [I] at a series of [I] values at each of two substrate levels. Where the lines cross, the values of $1/v$ and [I] are the same for both substrate levels. Since we are considering competitive inhibition, for which V_{max} is independent of $[I]$, we can write the following expression, which holds *only* at the intersection point:

$$K_m \frac{1}{[S]_1} + K_m \frac{1}{[S]_1} \frac{[I]}{K_i} = K_m \frac{1}{[S]_2} + K_m \frac{1}{[S]_2} \frac{[I]}{K_i} \qquad (6\text{-}35)$$

This may be rewritten as

$$\frac{1}{[S]_1} \left(1 + \frac{[I]}{K_i}\right) = \frac{1}{[S]_2} \left(1 + \frac{[I]}{K_i}\right) \qquad (6\text{-}36)$$

Since $[S]_1 \neq [S]_2$, Eq. 6-36 can hold only if $[I] = -K_i$. A horizontal line drawn above the $[I]$ axis at the value $1/V_{max}$ should intersect at this same point for competitive inhibition (see Figure 6-3a and 6-3c).

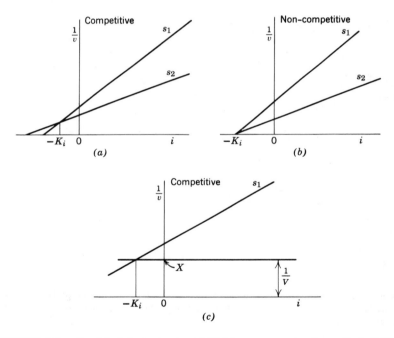

FIGURE 6-3. Graphical determination of inhibitor constant by the method of Dixon.

From the reciprocal expression for case c, it is seen that this method is also applicable to noncompetitive inhibition with one important difference—the intersection point is on the abscissa, that is, it occurs when $1/v = 0$. This can be seen by setting the Lineweaver-Burk equation for c equal to zero and solving for [I] (Fig. 6-3b). This method offers one of the clearest differentiations available between competitive and noncompetitive inhibition, but for the data to be reliable, one needs to ensure that the range of inhibition concentrations used spans the (assumed) value of K_i, or at least comes within 50% of it. Otherwise an excessive graphical extrapolation will be necessary.

References

1. M. Dixon and D. C. Webb, *Enzymes*, 2nd ed., Academic, New York, 1964, Chap. 6.
2. A. G. Ogston, *Nature*, **162**, 963 (1948).
3. G. Popjak, *The Enzymes*, vol. 2, 3rd ed., P. D. Boyer (Ed.), Academic, New York, 1970, p. 116.
4. M. L. Bender and F. J. Kezdy, *Ann. Rev. Biochem.*, **34**, 49 (1965).
5. M. Dixon, *Biochem. J.*, **55**, 170 (1953).

CHAPTER 7

Mechanisms of Catalysis

In the preceding chapters we discussed in some detail the principal techniques enzyme chemists use in their studies of enzymes and of enzyme-catalyzed reactions, but we scarcely touched upon the questions that are central to modern enzymology. How do enzymes work? By what chemical mechanisms do enzymes bring about their remarkable and specific rate accelerations? In what way, if any, is enzyme catalysis related to ordinary nonenzymatic catalysis? Can enzyme catalysis be understood in terms of the accepted concepts of physical organic chemistry, or must we seek understanding in terms of some special, new chemical principles that have no close parallel in the nonbiological realm?

It must be conceded at the outset that we cannot yet offer fully satisfactory answers to these questions and others related to them for one enzyme, let alone for enzymes in general. Although we do not in any case know, in detail or quantitatively, the mechanisms or driving forces that account for specific rate accelerations due to enzyme catalysis, and although there may in fact be important chemical principles involved in enzyme catalysis that we do not yet fully understand, it is apparent that at least to a very significant extent enzyme catalysis is indeed amenable to study using the techniques of physical organic chemistry and that understanding may be approached in terms of phenomena that have been observed in the behavior of relatively simple, nonbiological organic model systems. In other words, there appears to be no necessity to suggest that enzyme action is fundamentally different from any other chemical reaction.

Jencks[1] has pointed out that the problem of mechanism of enzyme action may be approached in three ways: (1) by theorizing; (2) by examining the properties of enzymes themselves, enzyme-substrate complexes, and enzyme-catalyzed reactions; and (3) by examining the nature of nonenzymatic reactions which are related to enzyme-catalyzed reactions and the means by which these reactions may be accelerated. Pure theory can never in itself provide the answers we seek but serves the valuable role of evaluating our experiments and stimulating new ones. The ultimate answers must of course come from the second approach, and in subsequent chapters we will look at the present state of progress along these lines for some of the more extensively studied enzymes. In this chapter we will, for the most part, adopt the first and third approaches on the premise stated earlier that enzyme action is in principle no different from any other chemical reaction and that enzymes operate by combining or integrating, in some specific and perhaps synergistic manner, elements of catalysis that may be observed and understood individually in appropriately designed model systems.

We will examine four mechanisms of catalysis which, conceptually at least, one might expect an enzyme to employ, individually or in concert, to bring about rate accelerations. These include the so-called approximation of reactants, covalent nucleophilic catalysis, general acid-base catalysis, and catalysis by induction of strain.

One more point remains to be considered before we examine these four mechanisms of catalysis. In proposing any chemical mechanism one must begin with a quantitative estimate of just what it is he is setting out to explain. The term "catalysis" in itself implies rate acceleration, so our starting point in elucidating mechanisms of catalysis must be measured rate enhancements. That is, just how much faster does a given reaction go under certain conditions in the presence of the catalyst than it goes under the same conditions in the catalyst's absence? All of this might at first seem trivial and rather obvious, which in fact it is, except for the troublesome fact that often a reaction which proceeds at a measurable rate under the influence of a catalyst may *not* proceed at a measurable rate (if at all) under comparable conditions in the absence of the catalyst. Or, if the uncatalyzed reaction does take place, it may do so by a pathway totally unrelated to the pathway of the catalyzed reaction, so that a comparison of the two rates is not very useful. Of course in cases where the noncatalyzed reaction is not observable, one can set a minimum estimate on the presumed catalytic rate acceleration by specifying the slowest noncatalyzed rate which his techniques would be capable of measuring. But clearly, one's conclusions are most valuable when he is dealing with systems in which direct comparisons between catalyzed and uncatalyzed rates are possible, and our discussion will concentrate on such cases.

Approximation of Reactants and Orientation Effects

Given our current conception of an enzyme-active site as a more or less rigidly defined region at the surface of an enzyme molecule which can reversibly bind substrate molecules in the manner of a template, an obvious means by which enzymes might effect rate accelerations is simply to position the reactants side by side. This should make the reaction somewhat more probable than it would be if it depended upon random collisions in (dilute) solution, especially if we confer on the enzyme the ability not only to bring the reactants close together at the active site but also to *orient* them geometrically, one with respect to the other, so that they may collide in the "right way."*

One approach to evaluating the possible contribution of proximity and orientation effects to rate accelerations is the comparison of the rates of analogous intermolecular and intramolecular reactions. Koshland[2,3] has pointed out that for a pair of reactions such as that indicated in Eqs. 7-1 and 7-2, a k_1/k_2 ratio of only about 5–50M can be anticipated from proximity effects alone. This conclusion may be arrived at as follows.

$$\begin{matrix} \lceil A \\ \\ \lfloor B \end{matrix} \xrightarrow{k_1 \, \text{sec}^{-1}} \begin{matrix} \lceil A \\ | \\ \lfloor B \end{matrix} \qquad\qquad (7\text{-}1)$$

$$A + B \xrightarrow{k_2 M^{-1} \text{sec}^{-1}} A\!-\!B \qquad\qquad (7\text{-}2)$$

Let us assume that the intrinsic reactivity of A with respect to B is the same in both cases, and that the first-order rate constant, k_1, is larger than the second-order rate constant, k_2, for the sole reason that in the intramolecular case (Eq. 7-1), B *always* has a nearest neighbor A group, whereas in the intermolecular case (Eq. 7-2), B must encounter A molecules at random. At a given instant only a fraction of the B molecules will have nearest neighbor A molecules and the magnitude of that fraction depends upon the *concentration* of A. Clearly, if B is kept dilute, it should in principle

---

* Consider, for example, a simple S_N2 displacement reaction. The attacking nucleophile must approach the substrate more or less directly from behind the carbon which bears

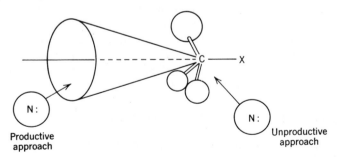

Productive
approach

Unproductive
approach

the leaving group. At least for purposes of discussion we can define the region of productive approach of the nucleophile as a cone whose axis corresponds with the bond joining the leaving group and the attacked carbon atom. Any nucleophile that approaches the substrate molecule outside this conical region of space will collide unproductively.

be possible to increase the concentration of A to a point where every B has, on the average, just one nearest neighbor A molecule. Let us call this the critical concentration of A or $[A]_{crit}$. Under such conditions (large excess of A over B) the intermolecular reaction (Eq. 7-2) will follow a pseudo-first-order rate law:

$$v_2 = k_{obs}[B] \qquad (k_{obs} \equiv k_2[A]_{crit}) \qquad (7\text{-}3)$$

Given our original assumptions, since every B now has a nearest neighbor A molecule, the two first-order rate constants k_1 and k_{obs} should be equal. Thus we have

$$k_1 = k_{obs} \equiv k_2[A]_{crit} \qquad (7\text{-}4)$$

or

$$\frac{k_1}{k_2} = [A]_{crit} \qquad (7\text{-}5)$$

Equation 7-5 says that the k_1/k_2 ratio is just the critical molar concentration of A necessary to provide every B molecule in Eq. 7-2 with a nearest neighbor A molecule. It should be apparent that the absolute upper limit for $[A]_{crit}$ (and thus for k_1/k_2 in cases where reactant approximation is the only factor favoring Eq. 7-1 over Eq. 7-2) is the concentration of A in pure liquid A. For example, if A were water, k_1/k_2 could be no greater than $55.5M$. In most cases, simple proximity effects must be somewhat smaller than this. For example, if A and B are both water-sized molecules, a dilute solution of B in A will provide each B molecule with about 12 nearest neighbor A molecules.[2] Thus in a water solution with B dilute, $[A]_{crit}$ would be only $55.5/12$ or $4.6M$.

A k_1/k_2 ratio of about $5–50M$ is in fact often found when the rate of a simple uncatalyzed bimolecular reaction is compared with its intramolecular counterpart. Such is the case, for example, for the following comparison of the intramolecular versus intermolecular nucleophilic attack by imidazole on a p-nitrophenyl ester where k_1/k_2 is $23.9M$[7] (Eqs. 7-6 and 7-7). In a number of cases, however, such k_1/k_2 ratios greatly exceed this anticipated magnitude with values of about 10^3 to 10^8 not uncommonly observed.[1,4,5]

$$(7\text{-}6)$$

$$CH_3CO_2 \langle\!\!\!\bigcirc\!\!\!\rangle NO_2 + \underset{H}{\overset{\overset{\cdots}{N}}{\bigg[\!\!\!\!\diagdown\!\!\!\!\bigg]}}\xrightarrow{k_2} CH_3\overset{O}{\overset{\|}{C}}\!-\!N\!\diagup\!\!\!\!\diagdown\!\!\!\!N + HO\langle\!\!\!\bigcirc\!\!\!\rangle\!\!-\!NO_2$$

$$(7\text{-}7)$$

Consider, for illustration, the following two comparisons involving nucleophilic attacks on a carbonyl carbon by a tertiary amine[8] (Eqs. 7-8 and 7-9) and a carboxylate ion[9] (Eqs. 7-10 and 7-11), respectively. In these two cases, we are dealing with rate enhancements of the magnitude commonly observed in enzyme catalysis, rate enhancements which are by no means understandable in terms of the simple proximity effect ideas we have just considered. How are we to account for such enormous rate enhancement?

$$\overset{-CO_2C_6H_5}{\underset{-N(CH_3)_2}{\Big\langle}}\xrightarrow{k_1}\quad \overset{O}{\underset{-\overset{\oplus}{N}(CH_3)_2}{\Big\langle}} + C_6H_5O^{\ominus} \qquad (7\text{-}8)$$

$$CH_3CO_2C_6H_5 + (CH_3)_3N \xrightarrow{k_2} (CH_3)_3\overset{\oplus}{N}COCH_3 + C_6H_5O^{\ominus} \qquad (7\text{-}9)$$

$$\frac{k_1}{k_2} = 1.25 \times 10^3 M$$

$$(7\text{-}10)$$

$$\frac{k_1}{k_2} = 1.3 \times 10^8 M$$

$$CH_3CO_2\langle\!\!\!\bigcirc\!\!\!\rangle OCH_3 + CH_3CO_2^{\ominus} \xrightarrow{k_2} (CH_3CO)_2O + \langle\!\!\!\bigcirc\!\!\!\rangle \qquad (7\text{-}11)$$

Koshland recently suggested that in a neighboring-group reaction such as we have illustrated or in an enzyme-catalyzed reaction where the neighboring-group principle is in operation, rate enhancements vastly in excess of any that can be accounted for by proximity effects alone may be due to the operation of exceedingly sensitive orientation effects.[10] In discussing our earlier example of the orientation requirements of an S_N2 reaction, we noted that the nucleophile and the central carbon of the alkyl halide must approach each other on a line of collision that falls within a certain limiting conical region or solid angle with the carbon nucleus at its apex. Presumably the nucleophile itself may also have similar orientation requirements (i.e., it must enter the cone in a specific manner) so that we have the situation schematically illustrated in Figure 7-1.

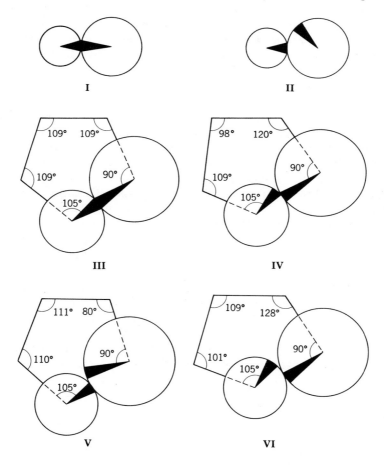

FIGURE 7-1. Possible orientation requirements for the productive approach of two reacting atoms. Structures I and II illustrate productive and unproductive collisions for atoms that must encounter each other at random. Structures III to VI represent cyclic or neighboring group situations where the collision orientation of the two atoms is fixed by the bond lengths and bond angles of nonreacting atoms. In III and IV, productive encounter can readily occur with a minimum of bond deformation in the cyclic systems. In V and VI, productive encounter would require much more extreme deformations, of bond lengths and bond angles.[10]

Only a small fraction of all possible collision orientation combinations of the two reacting atoms can lead to reaction. The reciprocal of this fraction, which Koshland designates θ, represents the orientational rate enhancement that might be achieved in a neighboring-group reaction or enzyme-catalyzed reaction where collision orientation is optimized. These orientation factors, θ, had always been assumed to range from unity up to 10^1 or 10^2; that is, of the same order of magnitude as proximity effects.[6] But Koshland claims,[10] on the basis of certain intramolecular versus intermolecular esterification and lactonization rate enhancements he has observed, that bimolecular θ factors may in fact range from 10^3 to 10^5. This presumed ability of enzymes to achieve enormous rate enhancements by precise alignment of reacting atoms has been termed *orbital steering*.

The orbital steering theory has been severely challenged by Bruice,[11] who points out that to account for rate enhancements of the order of 10^8 for certain intramolecular reactions he has observed (see, for example, Eqs. 7-11 and 7-12), θ factors of about 10^6 to 10^7 would have to be invoked after correcting for simple proximity effects. It is argued that to attribute such enormous rate enhancement factors to the precise alignment of reacting orbitals one must assume that covalent bonds are about 100 times more resistant toward minor bending deformations than molecular orbital theory and spectroscopic data indicate them to be. In other words, bond strength data suggest that enhancement factors of 10^6 to 10^8 cannot be achieved by orbital steering alone.

Another dimension in the rationalization of the exceptional catalytic powers of enzymes has been put forward by Reuben.[12] He points out that a bimolecular reaction may be described by the potential energy diagram of Figure 7-2 where A and B come together at a diffusion-controlled rate to form a fleeting AB complex which may then proceed to form the activated complex C^* or dissociate again to give solvated reactants. The time constant character-

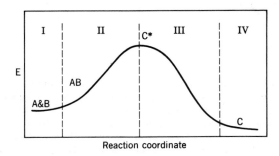

FIGURE 7-2. The potential energy diagram for a bimolecular reaction.

izing this AB complex is probably of the order of only 10^{-13} second. On the other hand, the mean lifetime of an enzyme-substrate complex, ABEnz, may very well be anywhere from 10^{-7} to 10^{-4} second. Thus the enzyme keeps the reactants in close proximity (i.e., in region II of the potential energy diagram) much longer than they would be in its absence, and this in turn markedly enhances the probability that the activated complex, C*, will be reached. It is claimed that this so-called *substrate anchoring* property of enzymes could account for rate enhancements of between 10^6 and 10^9.

As is the case with orbital steering, substrate anchoring is an attractive concept that may be at least qualitatively valid and is quite useful in the sense that it helps us grasp intuitively a factor that may indeed be involved in enzyme catalysis. On the other hand, it seems doubtful that such theories, as they are currently presented, could ever be established as correct or disproved as false on quantitative experimental grounds for individual enzymes, however conceptually valid they may be. There are factors involved in many enzyme catalyses that are more subject to experimental characterization and measurement, and we now turn our attention to these.

Covalent Catalysis by Nucleophiles

One of the more dramatic advances toward putting enzyme catalysis on firm mechanistic grounds has been the discovery that certain enzymes react with their substrates in the process of catalytic turnover to form discreet covalent enzyme-substrate intermediates. These covalent intermediates, some of which may actually be isolated and characterized, imply that in at least some cases, enzymes contribute more to catalysis than the ability to hold substrates in positions conducive to reaction. Moreover, it should be possible in these cases to identify with certainty at least one of the active-site functional groups of the enzyme in question—the one which participates in the covalent bond to substrate—and to offer some reasonable assessment of its role in the catalysis.

The initial task to be undertaken in a case where the involvement of a covalent intermediate is suspected is of course the gathering of evidence that *demonstrates* its involvement and elucidates its chemistry. It is beyond the scope of the present discussion to enumerate the approaches and techniques that have been, or might be, brought to bear on this problem. We will simply state here that the following criteria constitute *proof* that a certain covalent enzyme-substrate intermediate lies on the pathway of an enzymic reaction:

1. The isolation and chemical characterization of an intermediate.
2. The demonstration that this intermediate is formed and reacts further with rate constants which are consistent with the observed overall rate of the enzymic reaction.

TABLE 7-1 Some Enzymes That Give Rise to Covalent Intermediates

Enzyme Class and Examples	Covalent Intermediate	Reacting Group			
		Substrate	Enzyme	Coenzyme	Reference[a]
Serine proteases: trypsin, chymotrypsin, subtilisin, thrombin, elastase	Acyl-serine	Acyl moiety of amide or ester	Serine-β-OH	—	1
Sulfhydryl proteases: papain, ficin, bromelain	Acyl-cysteine	Acyl moiety of amide or ester	Cysteine-β-SH	—	1
Other sulfhydryl enzymes: cytidine triphosphate synthetase, glyceraldehyde-3-phosphate dehydrogenase	Acyl-cysteine	Carbonyl group	Cysteine-β-SH	—	2, 3
Phosphate transfer: phosphoglucomutase, alkaline phosphatase	Phosphoserine	Phosphate moiety of phosphate ester	Serine-β-OH	—	4, 5
Phosphate transfer: phosphoenolpyruvate-hexose transphosphorylase, succinic thiokinase	Phosphoryl-imidazole	Phosphate moiety of phosphate ester	Tertiary nitrogen of histidine imidazole ring	—	6, 7
Mammalian aldolases and decarboxylases	Schiff base (imine)	Carbonyl group	Lysine-NH_2	—	8
Pyridoxal phosphate	Schiff base	Amino group	Amino group	Aldehyde carbonyl	8

[a] References on facing page.

108

3. The demonstration that the same or a completely analogous intermediate is formed with "normal" substrates and under "normal" conditions in cases where the first two criteria can be met only with relatively poor "abnormal" substrates and/or under "abnormal" conditions.

Rarely, if ever, have all these criteria been fully satisfied. Many of the (tentative) conclusions which have been drawn concerning covalent enzyme-substrate intermediates rest upon extensive indirect evidence (usually kinetics), the nature of which we cannot discuss here[13] (see, however, Chapters 5 and 9).

Some of the enzymes for which enzyme-substrate intermediates have been implicated with varying degrees of certainty are listed in Table 7-1. Among these, the most intensively investigated examples are the esterases and proteases in which a nucleophilic amino acid side chain at the enzyme active site serves as an acceptor for the acyl portion of an ester or peptide substrate and subsequently transfers this acyl group to a second nucleophilic acceptor (often water) to complete the reaction. Phosphoryl-enzyme intermediates have been implicated in the enzymic action of several phosphate-transferring enzymes. Finally, Schiff base intermediates have been found to play a role in the operation of certain aldolases, decarboxylases, and enzymes employing pyridoxal phosphate as a cofactor (see Chapter 8).

The mechanism of action of some of these enzymes will be discussed in detail in later chapters of this book. Our primary concerns in the present discussion are (1) the advantage, so far as catalysis is concerned, for a reaction to proceed through a covalent intermediate rather than directly, and (2) what is known about covalent catalysis in simple systems that might serve as models for such behavior in enzymic reactions.

a. Acyl Transfer

We have noted that acyl transfer enzymes (esterases and proteases primarily) provide many of the best documented cases in which covalent intermediates between the enzyme and (a portion of) the substrate have been implicated. Nonenzymatic acyl transfer has been widely studied. The literature

1. M. L. Bender and F. J. Kezdy, *Ann. Rev. Biochem.*, **34**, 49 (1965).
2. A. Levitzky and D. E. Koshland, Jr., *Biochemistry*, **10**, 3365 (1971).
3. E. L. Taylor, B. P. Meriwether, and J. H. Park, *J. Biol. Chem.*, **238**, 734 (1963).
4. S. Harshman and V. A. Najjar, *Biochemistry*, **4**, 2526 (1965).
5. L. Engstrom, *Biochim. Biophys. Acta*, **52**, 49 (1961); **54**, 179 (1961); **56**, 606 (1962).
6. W. Kundig, S. Ghosh, and S. Roseman, *Proc. Nat. Acad. Sci. U.S.*, **52**, 1067 (1964).
7. G. Kreil and P. D. Boyer, *Biochem. Biophys. Res. Commun.*, **16**, 551 (1964).
8. E. E. Snell and S. J. DiMari, *The Enzymes*, Vol. 2, 3rd. ed, P. D. Boyer (Ed.), Academic, New York, 1970, Chap. 7.

provides a wealth of information concerning structure-reactivity relationships and mechanisms of catalysis for reactions of this type.[14] In view of this extensive literature on the subject, it comes as something of a disappointment to find very few unambiguous observations of catalysis of acyl transfer by nucleophiles, which is of course what we seek as a model for this type of enzyme action.

The following generalized equations illustrate the simplest possible mechanistic scheme for nucleophilic catalysis of acyl transfer and will serve as a point of departure as we attempt to understand why this phenomenon is not very common. The symbols L, A, and N are used to indicate the original leaving group, the final acyl group acceptor, and the nucleophilic catalyst,

Uncatalyzed:

$$
\begin{array}{c}
\overset{O}{\overset{\|}{R-C-L}} \underset{k_{-1}[L^-]}{\overset{k_1[A]}{\rightleftharpoons}} \left[\overset{O}{\overset{\|}{R-C-A^{\oplus}}}\right]
\end{array}
\tag{7-12}
$$

Catalyzed:

$$
\overset{O}{\overset{\|}{R-C-L}} \underset{k_{-2}[L^-]}{\overset{k_2[N]}{\rightleftharpoons}} \left[\overset{O}{\overset{\|}{R-C-N}}\right]^{\oplus} \overset{k_3[A]}{\longrightarrow} \left[\overset{O}{\overset{\|}{R-C-A}}\right]^{\oplus}
\tag{7-13}
$$

respectively. In order to observe catalysis, it is of course necessary that both steps of Eq. 7-13 be faster than Eq. 7-12. This requires first that the catalyst N must have a higher nucleophilic reactivity toward the original substrate than does the final acyl-group acceptor A. Second, the intermediate in Eq. 7-13 must be more reactive toward the acceptor than the substrate is. And finally, this intermediate must be thermodynamically less stable than the final product or it will accumulate at the expense of the product and tie up the catalyst.

It should be appreciated at once that these requirements imply a catalyst with some unusual properties, for it is often to be expected that the stronger the nucleophile, the more stable and less suceptible to a second nucleophilic attack will be the substitution product. In other words, good nucleophiles are commonly expected to be poor leaving groups. For example, in the hydrolysis of an ester (transfer of an acyl group from an alcohol to water) one would scarcely expect a primary amine to perform as a nucleophilic catalyst, for although the amine is more nucleophilic than water toward esters, the resultant amide is in general less reactive toward water than the original ester was and would tend to accumulate.

The classic case of nucleophilic catalysis of acyl transfer by an organic nucleophile is the catalysis of p-nitrophenyl acetate hydrolysis by imidazole.[15] The reaction involves the relatively slow attack by imidazole on the carbonyl carbon of the ester with the displacement of p-nitrophenolate. (Although

$$(7\text{-}14)$$

"slow" relative to the second step of the catalysis, imidazole attack is still much faster than direct attack by water on the initial ester.) Then water rapidly hydrolyzes the protonated acetylimidazole intermediate giving acetic acid and regenerating the catalyst.

This particular demonstration of nucleophilic catalysis by imidazole is especially noteworthy due to the frequently observed apparent involvement of the imidazolyl group of a histidine residue at the active site in the action of a number of hydrolytic enzymes (e.g., chymotrypsin, trypsin, and ribonuclease). But it appears, in fact, that the role of imidazole in such cases is not that of a nucleophile but of a general base (proton acceptor). We will examine acid-general base catalysis and discuss this different role of imidazole as a catalyst later. For the purposes of the present discussion, we need only note that imidazole is sufficiently nucleophilic to effect the type of catalysis observed in Eq. 7-14 only when the leaving group is a particularly labile one. Thus whereas nucleophilic catalysis is clearly observable when the weakly basic p-nitrophenolate ion is the leaving group, no such catalysis is observed when the potential leaving group would be strongly basic as is the case with an ordinary alkoxide leaving group. Although the unusually high reactivity toward nucleophilic displacement of N-acylimidazole intermediates* makes nucleophilic catalysis by imidazole attractive from one standpoint, the fact that the imidazole nitrogen is only moderately nucleophilic limits such catalysis to unusually reactive "energy-rich" acyl substrates.

* This high reactivity is due primarily to the fact that the N-acylimidazole is easily protonated relative to ordinary amides and to the fact that the amide bond has very little resonance stabilization due to the participation of the nitrogen lone pair in the aromatic π-system of the imidazole ring.

The nucleophiles that have actually been identified as giving rise to acyl-enzyme intermediates in esterases and proteases are the side chain oxygen of serine and the corresponding sulfur of cysteine (see Table 7-1). Although no examples of nonenzymatic nucleophilic catalysis of acyl transfer by serine or cysteine derivatives (or by any other alcohols, alkoxides, thiols, or thiolates) have yet been reported, there is some evidence that they may be suited to this function under some conditions. For example, N-acetylserinamide has a nucleophilic reactivity toward base-catalyzed acylation by energy-rich acyl compounds (acetylimidazole and *p*-nitrophenylacetate) some three orders of magnitude higher than the reactivity of water or ethanol.[16] Moreover, the product of such an acylation, N,O-diacetylserinamide, is anomalously reactive toward general base-catalyzed deacylation by amine nucleophiles, though not by water.[16] Thiols in general are substantially more nucleophilic than alcohols or nitrogen bases, so in that respect the cysteine sulfhydryl group might be expected to perform well as a nucleophilic catalyst. On the other hand, thiol esters are no more reactive toward water or hydroxide ion than are normal oxygen esters, though they may in fact be more reactive than oxygen esters toward some other nucleophiles.

b. Phosphate Transfer

Table 7-1 indicates that active-site nucleophiles might contribute in a catalytic way not only to enzymic acyl transfer but also to enzyme-catalyzed phosphate transfer. Bimolecular nucleophilic catalysis in nonenzymic phosphate transfer reactions has in fact been observed.[17] The cases reported involved the catalysis of hydrolysis of acetyl phosphate by several tertiary amines, including triethylenediamine, pyridine, and 4-methylpyridine.

In the case of triethylenediamine, the unstable intermediate phosphorylated amine can effect phosphate transfer not only to the solvent water but also to fluoride ion, which competes with water in nucleophilic attack on phosphorous. This demonstrates that the catalysis is indeed nucleophilic catalysis rather than general base catalysis.* A mechanism consistent with these observations is given in Eq. 7-15.

$$(7\text{-}15)$$

* General base catalysis would generate no intermediate for a competing nucleophile to attack.

Unfortunately, it appears that such behavior is not very common. Other amines (among them, aniline, ammonia, n-butylamine, glycine, and notably imidazole and N-methylimidazole) are *general base* catalysts for acetylphosphate hydrolysis and give C—O bond fission instead of the P—O bond fission observed with nucleophilic catalysis. No good explanation for this difference in behavior among several amines has yet been offered. Furthermore, oxygen and sulfur nucleophiles are apparently without effect on the hydrolysis of *mono* and *di*substituted phosphates. In fact, aside from the case mentioned above, nonenzymic intermolecular nucleophilic catalysis of phosphate transfer appears not to have been observed. Some cases of possible intramolecular nucleophilic attack on substituted phosphates have been recorded,[18] but even in these cases the situation is far from straightforward.

It appears in general, at least for monosubstituted phosphates, that (1) the phosphorus is not very susceptible to nucleophilic attack, (2) various nucleophilic reagents differ very little in reactivity toward such compounds, and (3) various potential phosphorylated "intermediates" that might presumably be involved in phosphate transfer reactions show little selectivity toward different phosphoryl acceptors. Thus although phosphorylated enzymes have been detected or isolated as indicated in Table 7-1, the behavior of the nonenzymic model systems which have been investigated would suggest that nucleophilic catalysis per se is less likely in phosphate transfer than it is in acyl transfer.

c. Catalysis Involving Carbonyl Schiff Bases

The final category of covalent enzyme-substrate intermediates included in Table 7-1 is the Schiff base, which may be formed between a primary amino group of the enzyme and a carbonyl carbon of a substrate or coenzyme. A number of examples of catalysis of nonenzymatic reactions of carbonyl compounds involving Schiff base (or imine) intermediates have been observed and there is every reason to believe that what is understood about these model systems can be readily extended to analogous enzyme-catalyzed cases.

Consider first the decarboxylation of acetoacetic acid. The reaction is facilitated by the electron-withdrawing nature of the β-carbonyl group and presumably proceeds via the mechanistic pathway shown (Eq. 7-16).[14] Whether the proton in the electrically neutral reactant is best represented as belonging to the carbonyl oxygen (as in the zwitterion) or to the carboxyl oxygen is a moot point. In any case, the reaction is facilitated by the protonation of the carbonyl oxygen, which makes the carbonyl group a more effective electron sink.

This decarboxylation reaction is subject to covalent catalysis by amines through a mechanism which probably involves a Schiff base intermediate

$$(7\text{-}16)$$

(Eq. 7-17).* The observation that the cyanomethylamine-catalyzed decarboxylation of acetoacetic acid in aqueous solution proceeds at the same rate as the reaction of cyanomethylamine with ethyl acetoacetate to give a Schiff base strongly indicates that the Schiff base is indeed involved in the decarboxylation and that its formation is rate determining.[20] The observed catalysis results from the fact that the Schiff base nitrogen is much more readily protonated than the carbonyl oxygen, and thereby is a more effective electron sink in the decarboxylation.

$$(7\text{-}17)$$

* Although it is highly likely that the reaction proceeds through the protonated Schiff base as shown, this has not been conclusively demonstrated.

The mechanism of amine-catalyzed nonenzymatic decarboxylation of acetoacetic acid just discussed is almost certainly closely paralleled by enzymic decarboxylation by acetoacetate decarboxylase.[21] The catalytic amine in the enzyme is the ε-amino group of an active-site lysine residue. The Schiff base which this lysine forms with acetoacetate can be trapped (after decarboxylation) by reduction with sodium borohydride (Eq. 7-18). This reduced enzyme-substrate intermediate is totally inactive and upon total hydrolysis yields an amino acid mixture from which N-isopropyllysine can be isolated.

$$
\text{E—lys—NH}_2 + \text{CH}_3\overset{\text{O}}{\overset{\|}{\text{C}}}\text{CH}_2\overset{\text{O}}{\overset{\|}{\text{C}}}\text{OH} \longrightarrow
$$

$$
\text{E—lys—}\overset{\oplus}{\text{NH}}{=}\text{C}\underset{\text{CH}_3}{\text{CH}_2}\overset{\text{O}}{\overset{\|}{\text{C}}}\text{OH} \xrightarrow{-\text{CO}_2} \text{E—lys—N}{=}\underset{\text{CH}_3}{\text{C}}\text{—CH}_3 \xrightarrow{\text{NaBH}_4}
$$

$$
\text{E—lys—NH—CH—CH}_3 \quad\underset{\text{CH}_3}{} \xrightarrow{\text{hydrol.}} \text{lys—NH—CH(CH}_3)_2 \quad (7\text{-}18)
$$

Aldol condensations are also subject to amine catalysis.[22] Equation 7-19 illustrates a likely mechanism for the amine catalysis of the dealdolization (aldol condensation in reverse) of diacetone alcohol. Here again, as in the

$$
\underset{\text{CH}_3}{\overset{\text{OH}}{\text{CH}_3\text{C}}}\text{—CH}_2\overset{\text{O}}{\overset{\|}{\text{C}}}\text{CH}_3 + \text{RNH}_2 \rightleftharpoons \underset{\text{CH}_3}{\overset{\text{OH}}{\text{CH}_3\text{C}}}\text{—CH}_2\underset{\underset{\oplus}{\text{H}_2\text{NR}}}{\overset{\text{O}^\ominus}{\text{C}}}\text{CH}_3 \rightleftharpoons
$$

$$
\underset{\underset{\oplus}{\text{CH}_3 \quad \text{H}_2\text{NR}}}{\overset{\text{O}^\ominus \qquad \text{OH}}{\text{CH}_3\text{C}\text{—CH}_2\text{C}\text{CH}_3}} \xrightarrow{\pm\text{H}_2\text{O}} \underset{\underset{\underset{\oplus}{\text{R} \quad \text{H}}}{\text{N}}}{\overset{\text{O}^\ominus}{\text{CH}_3\text{C}\text{—CH}_2\text{C}\text{—CH}_3}}\text{—} \rightleftharpoons \underset{\text{CH}_3}{\overset{\text{O}}{\overset{\|}{\text{CH}_3\text{C}}}} + \underset{\underset{\text{R} \quad \text{H}}{\text{N:}}}{\text{CH}_2{=}\text{C—CH}_3}
$$

$$
(7\text{-}19)
$$

$$
\underset{\underset{\text{R} \quad \text{H}}{\text{N}}}{\text{CH}_2{=}\text{C—CH}_3} \rightleftharpoons \underset{\underset{\text{R}}{\text{N}}}{\text{CH}_3\text{—C—CH}_3} \xrightarrow{\pm\text{H}_2\text{O}} \text{CH}_3\overset{\text{O}}{\overset{\|}{\text{C}}}\text{—CH}_3 + \text{RNH}_2
$$

case of amine-catalyzed decarboxylation of β-keto acids, conversions of the carbonyl group to a protonated Schiff base provides an effective electron sink for the difficult step of the reaction, which is carbon-carbon bond cleavage.

In the condensation direction we have Eq. 7-19 in reverse. The initially formed acetone Schiff base is in tautomeric equilibrium with the corresponding enamine; an equilibrium which favors enamine more than the analogous keto-enol tautomerization favors enol. The enamine itself is a resonance hybrid with considerable electron density on carbon and may in fact be regarded as a stabilized carbanion, which quite readily attacks a carbonyl group to form the new carbon-carbon bond.

$$CH_3-\underset{\underset{R}{\overset{|}{\underset{\displaystyle N:}{}}}\diagdown H}{C}=CH_2 \longleftrightarrow CH_3-\underset{\underset{R}{\overset{\|}{\underset{\displaystyle N}{}}}\overset{\oplus}{\diagdown} H}{C}-CH_2:^{\ominus}$$

As with enzymic β-keto acid decarboxylation, there is strong evidence that the mammalian aldolase enzymes which catalyze aldol condensations and dealdolizations employ amine catalysis essentially like that observed for the nonenzymic models (see Chapter 11). Both mammalian aldolases and transaldolases are irreversibly inhibited when exposed to substrate in the presence of sodium borohydride. If the substrate is radioactive, this irreversible inhibition results in the incorporation of the radioactive label into the protein.[23] Furthermore, total acid hydrolysis of the inactivated enzymes gives rise to the expected N-ε-lysine derivatives,[24] just as has already been described for acetoacetate decarboxylase.

Amine catalysis through Schiff base formation is evidently important in many other enzymic reactions of carbonyl compounds, particularly those in which pyridoxal phosphate is involved as a coenzyme (see Chapter 8). Many of the reactions involved in amino acid metabolism require as a first step the formation of a Schiff base between the amino group of the substrate and the carbonyl group of the required coenzyme, pyridoxal phosphate. It is significant to note that this coenzyme generally is already present in the form of a Schiff base with the ε-amino group of a lysine residue of the enzyme in question. This is catalytically advantageous since it has been demonstrated in many cases that the amine functionality will attack a cationic Schiff base in a *trans*-imination reaction more readily than it will attack a free carbonyl group. For example, aniline catalyzes semicarbazone formation with pyridoxal aldehydes[25] (Eq. 7-20), and semicarbazone formation with pyridoxal phosphate itself is subject to catalysis by both primary and secondary amines[26] (Eq. 7-21).

$$(7\text{-}20)$$

$$(7\text{-}21)$$

General Acid-Base Catalysis

Many of the reactions we are concerned with in this book—indeed many reactions in organic chemistry—require proton transfers to or from reactants as they are transformed to products. Thus catalysis by proton-transferring agents, that is, by acids and bases, is a very frequently encountered phenomenon which is certainly an important consideration in a great many enzymic reactions.

There are two categories of proton transfer catalysis that are distinguishable from one another on experimental grounds. On the one hand we have reactions whose rates vary linearly with lyate ion activities ($a_{H_3O^+}$ and a_{OH^-} when the solvent is water) but which are not affected by the nature or concentration of any other Brønsted acids or bases present in solution. Those reactions are said to be *specific acid* catalyzed or *specific base* catalyzed as the case may be. Figure 7-3a illustrates the behavior of a specific-base catalyzed reaction. On the other hand, we have reactions whose rates reflect the nature and concentrations of *all* acidic or basic species present, including the lyate species. The terms *general acid* or *general base* catalysis apply to these, and Figure 7-3b shows the experimental manifestations of such catalysis.

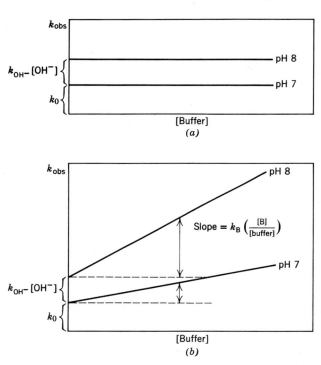

FIGURE 7-3. Specific base versus general base catalysis. (*a*) The observed reaction rate increases with pH but remains unaffected by the concentration of the conjugate base of a buffer. This is specific base catalysis and is characterized by the equation

$$k_{obs} = k_0 + k_{OH^-}[OH^-]$$

(*b*) The observed rate increases with pH as in (*a*) but the magnitude of the increase depends on the concentration of the conjugate base of the buffer. The rate law for this general base-catalyzed reaction is

$$k_{obs} = k_0 + k_{OH^-}[OH^-] + k_B[B]$$

Since enzymes operate optimally only in a relatively narrow pH range, usually near neutrality, our concern here is primarily with the phenomena of *general* acid-base catalysis. It is highly unlikely that any enzyme could bring about the significant increases in $a_{H_3O^+}$ or a_{OH^-} at the active site, which would be required to bring about effective *specific* acid or base catalysis in water near neutral pH. On the other hand, the presence at enzyme active sites of acidic and basic amino acid side chains which might serve as catalytic proton donors or acceptors near neutral pH is certainly common, if not virtually universal among enzymes which catalyze reactions in which proton transfers take place. Furthermore, many of these reactions, when carried out in the absence of enzymes, are indeed found to be subject to general acid-base catalysis. Among the general types of reactions of biochemical interest in this category are carbonyl additions and tautomerizations, hydrolyses and aminolyses of esters, and reactions of phosphates and polyphosphates.

a. The Brønsted Catalysis Law

The observed rate constant for a reaction which is subject to general acid and/or base catalysis is, as we have already implied, a function of the individual catalytic constants for each acid and/or base present and their concentrations:

$$k_{obs} = k_0 + \sum_{}^{i} k_{HA_i}[HA_i] + \sum_{}^{j} k_{B_j}[B_j] \tag{7-22}$$

These catalytic constants may be determined for any given catalytic species as indicated in Figure 7-3*b*. As might be expected, the strongest acids and bases are generally the most effective catalysts; indeed, if the logarithms of the catalytic constants for a number of catalysts of a given reaction are plotted against their respective pK_a values, a linear relationship is found* (Figure 7-4). This is the so-called Brønsted relationship and it is defined by one of the following Brønsted equations:

For general acid catalysis:

$$\log k_{HA} = C_A - \alpha(pK_a) \tag{7-23}$$

For general base catalysis:

$$\log k_B = C_B + \beta(pK_a) \tag{7-24}$$

where the slopes are given by α and β and where C_A and C_B are the intercepts on the ordinate.

The Brønsted coefficients α or β are thus a measure of the sensitivity of the reaction in question to the acidity or basicity of the catalysts employed.

---

* This statement is subject to two important limitations. First, the correlation may be expected to hold only if the catalysts being compared are of the same general type (e.g., all carboxylic acids or all primary amines). When comparing a variety of different catalysts, significant deviations are often observed as shown in Figure 7-4. There are a number of possible explanations for this, which we cannot discuss here.[27] Second, for catalysts that can donate or accept more than one proton, such as dicarboxylic acids or diamines, statistical corrections must be applied to the observed pK_a values.[27]

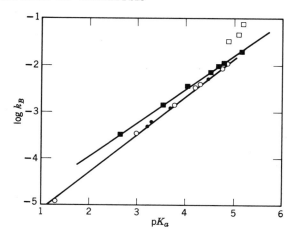

FIGURE 7–4. Brønsted plot of the catalytic constants k_B for the general-base catalyzed decomposition of nitramide. O = Anions of monocarboxylic acids; ● = singly-charged anions of dicarboxylic acids; ■ = ring-substituted anilines; □ = tertiary amines. Note the slightly different slopes for carboxylic ions (β = 0.82) and substituted anilines (β = 0.72). The tertiary amines deviate still further. (From the original data of J. N. Brønsted and K. Pederson, *Z. Physik. Chem.*, 108, 185 (1924) as presented by L. P. Hammett in *Physical Organic Chemistry*, 2nd ed., McGraw-Hill, New York, 1970, p. 318).

Values of α or β range between limits of zero and unity. Reactions with values of α or β approaching unity are extremely sensitive to acid or base strength, so much so that among all possible species in aqueous solution, only H_3O^+ or OH^- are strong enough to effect noticeable rate enhancements, even in situations where other acidic or basic species are present in vastly greater concentrations. Thus general acid or base catalysis with α or $\beta \simeq 1.0$ appears experimentally to be specific catalysis. On the other hand, if α or β approaches zero, the reaction is insensitive to catalyst acidity or basicity. For a reaction carried out in water, this means that water itself is the only effective catalyst since it can serve either as an acid or a base and is present in vast excess over any other possible catalyst. Thus a general acid- or base-catalyzed reaction with α or $\beta \simeq 0$ will appear experimentally to be uncatalyzed by acids or bases.

b. The Mechanistic Significance of General Acid-Base Catalysis and the Brønsted Parameters

The fact that concentration terms for general acids and bases appear in the rate equation for a reaction subject to general catalysis implies in itself that the transition state of such a reaction must formally include a proton and the conjugate bases of both the substrate and the catalyst (Eq. 7-25).

$$S + HA \longrightarrow \left[\overset{\delta+}{S}\text{---}H\text{---}\overset{\delta-}{A}\right]^{\ddagger} \longrightarrow \text{Products}$$

$$\text{(7-25)}$$

$$SH + B \longrightarrow \left[\overset{\delta-}{S}\text{---}H\text{---}\overset{\delta+}{B}\right]^{\ddagger} \longrightarrow \text{Products}$$

That is, general acid-base catalysis will not occur unless the presence of the catalyst in the transition state provides a more favorable pathway for reaction than would be possible in its absence.

Given this mechanistic interpretation of general acid-base catalysis, we next seek to understand what it is that the catalyst does to stabilize the transition state of which it is a necessary part. This understanding may be approached through a closer examination of the implications of the Brønsted catalysis laws. The logarithm of an acid dissociation constant is proportional to its free energy of dissociation:

$$\Delta G° = -RT \ln K_a = 2.303RT(pK_a) \qquad \text{(7-26)}$$

Moreover, transition state theory holds that the logarithm of a rate constant is proportional to the free energy of activation ($\Delta G^{\ddagger}$). Thus the linear relationship between pK_a and $\log k_{HA}$ in Eq. 7-23 or between pK_a and $\log k_B$ in Eq. 7-24 is actually a linear relationship between free energies for two different but evidently related phenomena involving a common family of chemical species, in this case Brønsted acids and bases. A number of such *linear free energy relationships*, among which the Brønsted catalysis laws were the first proposed, are now recognized and used by organic chemists[28] in their efforts to understand and predict trends in chemical properties resulting from structural changes in related compounds.

Since the free energy of acid dissociation ($\Delta G°$) is associated with the complete transfer of a proton from the acid to the standard base water, and since the free energy of activation ($\Delta G^{\ddagger}$) for general acid catalysis is associated with the partial transfer of a proton from the acid to a basic center in the substrate, it seems reasonable that the two processes should be similarly affected by changes in structure of the acid. Furthermore, it is often assumed that general acid or base catalysis may be attributed simply to the stabilization due to hydrogen bonding in the transition state, and that the Brønsted coefficient α or β is a measure of the extent to which proton transfer is complete in the transition state. Although these assumptions are probably correct to a very rough first approximation, the situation is clearly more complex than such assumptions indicate. The transition states associated with a given reaction in the presence and in the absence of a general acid or base catalyst differ by more than just the presence or absence of hydrogen bonding, because other bond-making and bond-breaking processes in addition to proton transfer are under way simultaneously and will be influenced by the presence of the catalyst.

c. Mechanism of General Acid-Base Catalysis

We have already established that the mechanistic feature which distinguishes general acid-base catalysis from specific acid-base catalysis is the involvement of the actual proton transfer in the transition state. If the proton is fully transferred in an equilibrium step before the transition state is reached, *specific* acid or base catalysis is observed because proton transfer equilibria in water depend only upon the pH. If the proton transfer begins only after the transition state is passed, no acid or base catalysis is observed.

The three mechanistic situations just described are illustrated below with three possible pathways for the reversible addition of water to a Schiff base (Eqs. 7-27 to 7-29). Inspection of the transition states for each of these three possibilities reveals that all of them are subject to the objection that they lead directly to the formation of high-energy oxonium ion intermediates. The "no-catalysis" case of Eq. 7-29 is particularly unattractive and might be eliminated a priori because of its exceedingly unlikely oxonium amide zwitterion intermediate.

Reactions such as this one are in fact found *experimentally* to be subject to general acid catalysis, which would appear to establish Eq. 7-28 as the

Specific Acid Catalysis

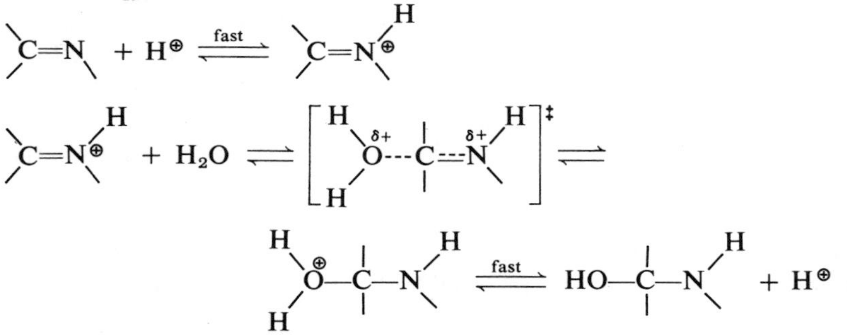

$$\text{(7-27)}$$

General Acid Catalysis

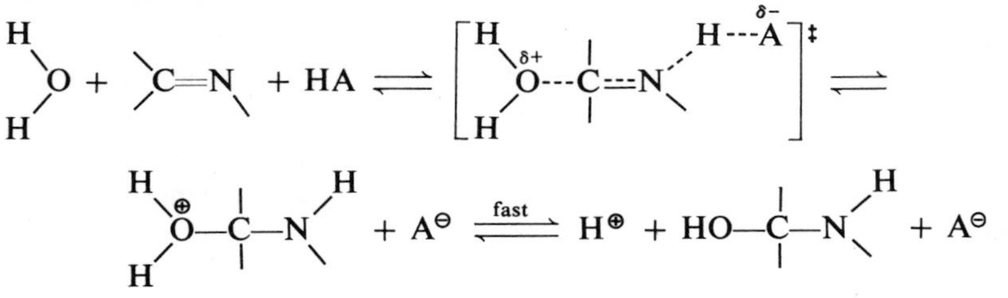

$$\text{(7-28)}$$

No Catalysis

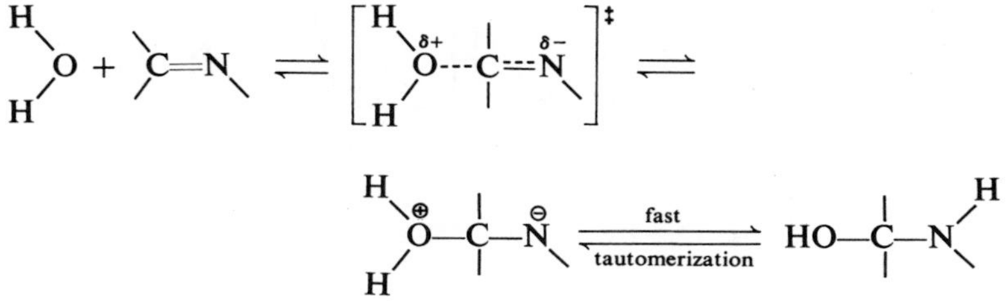

$$(7\text{-}29)$$

Specific Acid-General Base Catalysis

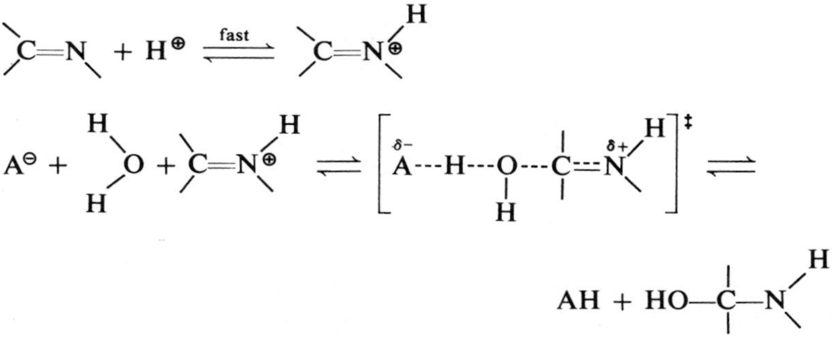

$$(7\text{-}30)$$

correct mechanism. However, an alternative mechanism (Eq. 7-30) can be formulated which involves a pre-equilibrium protonation of the imine nitrogen followed by general base catalysis of water attack on the electrophilic carbon atom. Unfortunately, true general acid catalysis (Eq. 7-28) and specific acid followed by general base catalysis (Eq. 7-30) give rise to identical rate laws and thus are kinetically indistinguishable. The rate law for Eq. 7-28 is

$$v = k[H_2O]\left[\,\rangle C{=}N{-}\,\right][HA] \tag{7-31}$$

and that for Eq. 7-30 is

$$v = k'[H_2O]\left[\,\rangle C{=}\overset{\oplus}{N}\overset{H}{\diagup}\,\right][A^-] \tag{7-32}$$

But since

$$K_{HA} = \frac{[H^+][A^-]}{[HA]}; \qquad [A^-] = \frac{K_{HA}[HA]}{[H^+]}$$

and

$$K_a' = \frac{[H^+]\left[\begin{array}{c}\diagdown\\ C=N-\\ \diagup\end{array}\right]}{\left[\begin{array}{c}\diagdown\quad\quad H\\ C=N^{\oplus}\\ \diagup\quad\quad\diagdown\end{array}\right]}; \qquad \left[\begin{array}{c}\diagdown\quad\quad H\\ C=N^{\oplus}\\ \diagup\quad\quad\diagdown\end{array}\right] = \frac{[H^+]\left[\begin{array}{c}\diagdown\\ C=N-\\ \diagup\end{array}\right]}{K_a'}$$

Eq. 7-32 becomes

$$v = k'[H_2O]\left(\frac{[H^+]\left[\begin{array}{c}\diagdown\\ C=N-\\ \diagup\end{array}\right]}{K_a'}\right)\left(\frac{K_{HA}[HA]}{[H^+]}\right)$$

$$v = k'\frac{K_{HA}}{K_a'}[H_2O]\left[\begin{array}{c}\diagdown\\ C=N-\\ \diagup\end{array}\right][HA] \tag{7-33}$$

which is identical in form to Eq. 7-31.

The ambiguity illustrated here by the kinetic equivalence of Eqs. 7-28 and 7-30 is very commonly encountered. It has to do with the fact that a rate law indicates the *stoichiometry* of a transition state but provides no information concerning how the components of that transition state are arranged structurally. In our example the transition states for both mechanisms consist formally of (1) the conjugate base of the substrate, (2) the conjugate base of the general catalyst, (3) a water molecule, and (4) a proton. The same kind of ambiguity exists with respect to true general base catalysis on the one hand and specific base followed by general acid catalysis on the other. Here again, both mechanisms give rise to the same form of rate law—that which manifests itself in experimental general base catalysis. A further generalization concerning mechanisms of this type is that whatever mechanism describes the reaction in the forward direction, the principle of microscopic reversibility requires that the reverse mechanism must be the opposite. Thus, for example, Eq. 7-30 is specific acid-general base catalyzed in the forward direction but, as shown, is subject to true general acid catalysis in reverse. Similarly, the reverse of Eq. 7-28 is an example of specific acid-general base catalysis.

In some cases it is indeed possible to design experiments that can resolve ambiguities of this type.[29] We will not concern ourselves with experiments of this kind here, but we will point out as a tentative generalization that general acid-base catalysis occurs in such a way as to avoid the unstable intermediates which are associated with alternate mechanistic pathways. Thus in our example, although Eq. 7-28 cannot be ruled out on any experimental grounds we have considered here, it is clearly less attractive than Eq. 7-30, in which no oxonium ion or similarly unstable intermediate appears.

A further implication of this generalization is that for facile reactions where little can be gained by general acid-base catalysis, such catalysis will not be important or may not be observable at all. For example, in the case of addition of nucleophiles to carbonyl functions, the importance of general acid catalysis varies as a function of the extent to which nucleophile-carbon bond making and concomitant carbon-oxygen bond breaking have proceeded in the transition state. For weak nucleophiles such as water, the transition state occurs very late in the sense that the bond-making and breaking processes are nearly complete, and in the transition state the carbonyl oxygen has acquired most of the electron density of the former π bond. In such a case, hydrogen bonding from a general acid catalyst to the highly developed charge on the carbonyl oxygen can lend considerable stability to the transition state, and in fact the attack of water on carbonyl functions is almost completely dependent on general acid catalysis. On the other hand, for very strongly basic amine nucleophiles, this reaction may exhibit no acid catalysis at all.[30] This is consistent with an early transition state where bond-making and breaking processes are barely under way and the presence in the transition state of a hydrogen bond donating catalyst would contribute little or no stabilization.

d. Concerted General Acid-Base Catalysis

Many reactions of biological interest require two or more proton transfers and it is attractive conceptually to postulate that an enzyme might very well effect simultaneous or concerted general acid-base catalysis at more than one position on a substrate. If so, this might help to account for the very large specific rate enhancements associated with enzyme catalysis. Consider the (strictly hypothetical) examples of Eqs. 7-34 to 7-37. In the absence of enzymes, each of these reactions is indeed found to be subject to catalysis

Hydration of an Aldehyde

$$(7\text{-}34)$$

Hydrolysis of a Peptide

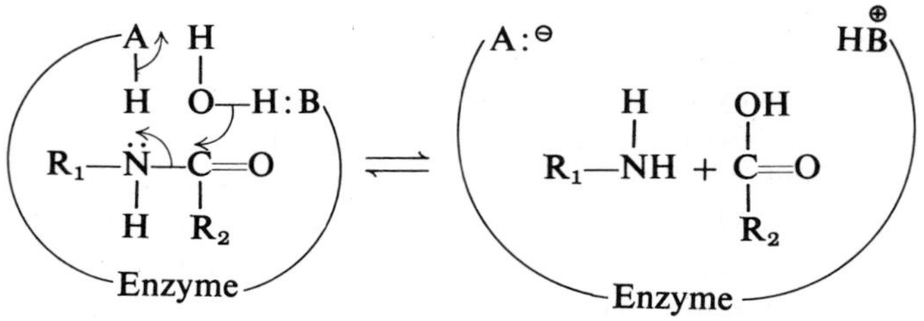

$$(7-35)$$

Mutarotation of Glucose

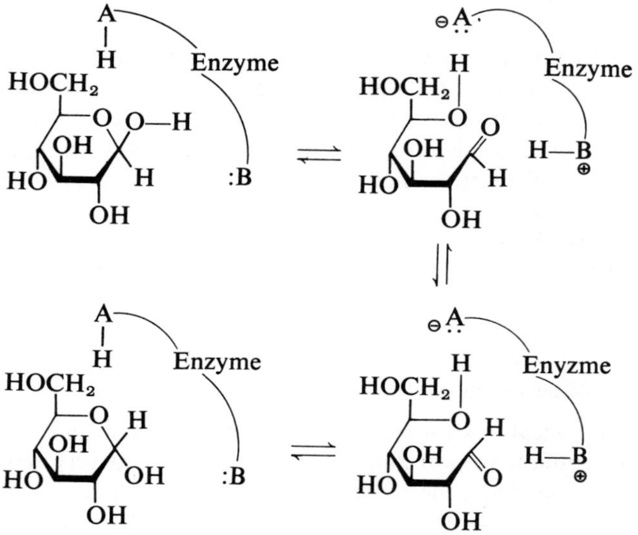

$$(7-36)$$

Tautomerization of a Carbonyl Compound

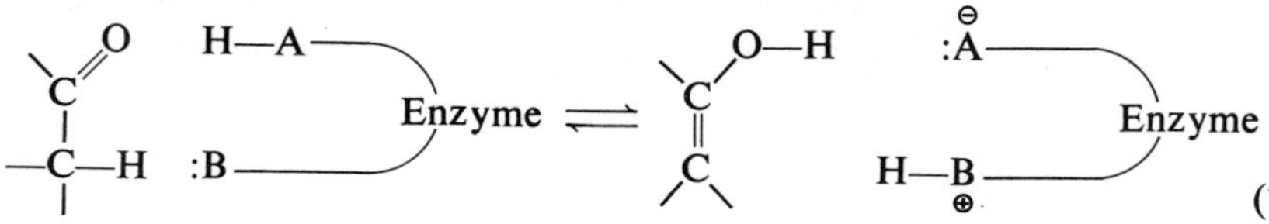

$$(7-37)$$

by both acids and bases. For example, in aqueous acetate buffers the mutarotation of glucose has a rate law of the form

$$k_{obs} = k_0 + k_H[H^+] + k_{OH}[OH^-] + k_{HA}[HOAc] + k_A\text{-}[OAc^-] \quad (7\text{-}38)$$

where the k_0 term for the "uncatalyzed" rate very likely reflects general acid and general base catalysis by water itself. But in order to support the idea that *concerted* general acid-base catalysis can be important for reactions such as this, we should like to find examples of reactions that have rate law terms such as $k[HA][B]$, indicating the simultaneous participation of both an acid and a base catalyst in the transition state.

In a classic study, Swain and Brown[31] found that the mutarotation of tetramethylglucose in the solvent benzene proceeds at an immeasurably slow rate, if at all, in the absence of any added acid or base catalyst. The reaction will occur in the presence of either phenol or pyridine, but the rate is still very slow. However if *both* pyridine and phenol are included in the reaction mixture, the mutarotation of tetramethylglucose in benzene proceeds quite rapidly with a rate law in which a third-order term overwhelmingly predominates (Eq. 7-39). Furthermore, it was found that α-pyridone, in which

$$v \simeq k[\text{Glu}][\text{pyridine}][\text{phenol}] \qquad (7\text{-}39)$$

a general acid and a general base are combined in the same molecule, catalyzes this reaction much more effectively than a mixture of phenol and pyridine (Eq. 7-40). At $0.05M$ catalyst concentration α-pyridone is 50 times

$$(7\text{-}40)$$

more effective than a phenol-pyridine mixture, and at $0.001M$ it is 7000 times more effective; all this despite the fact that the acidic and basic groups of α-pyridone are substantially *weaker* as acids and bases than phenol and pyridine, respectively!

Unfortunately, there are severe objections to the acceptance of the Swain-Brown study as a suitable model for concerted acid-base catalysis by enzymes. First, enzymes operate in water, not in benzene, and no one has ever found concerted catalysis in an aqueous medium which even distantly approaches the effectiveness of the Swain-Brown example. It can be argued that concerted catalysis in the mechanistic sense might indeed occur in water with a water molecule itself playing the role of one or both of the catalytic groups. This of course would not be evident in the kinetics. But such arguments are academic since an enzyme must catalyze a reaction faster than it occurs in water. Furthermore, the experiments with reactions in benzene solution have not been shown to demonstrate clear-cut concerted catalysis. Reactions with polar transition states in nonpolar solvents can be accelerated by several orders of magnitude in the presence of low concentrations of salts.[32] It is therefore entirely possible that much of the rate enhancement of tetramethylglucose mutarotation in the presence of acid-base pairs like phenol and pyridine is attributable to the presence of the ion pairs formed by a simple acid-base reaction. In fact it has been shown that several salts which could not be effective in concerted acid-base catalysis do indeed catalyze the mutarotation of tetramethylglucose in benzene.[33]

On the other hand, the objections raised to the unsuitability of benzene as a medium for enzyme-model studies may themselves be challenged on the grounds that although enzymes do indeed function in aqueous media, the microscopic environment provided by enzyme-active sites may in many cases actually be hydrophobic or "benzene-like." Indeed, Perutz has called particular attention to the fact that the one feature shared by all of the protein structures that have been elucidated by X-ray analysis is *"the exclusion of polar residues from the interior, except for special purposes connected with function."*[34] Furthermore, he has offered the intriguing suggestion that the ability of enzymes to provide electrostatic catalysis in a nonpolar environment within a bulk aqueous solvent might account in part for large rate enhancements. That is to say, in an enzyme active site which is hydrophobic in character, the presence of an ion pair or even a strong permanent dipole may contribute to catalysis in much the same way that salts in low concentration accelerate nonenzymic reactions in nonpolar solvents like benzene. Moreover, to the extent that experiments of the Swain-Brown type do indeed indicate the possible importance of concerted general acid-base catalysis in nonpolar media, it can be argued that such catalysis may very well be important as a factor in enzyme catalysis, at least in cases where the protein provides a hydrophobic environment for the reaction under consideration.

Catalysis by Induction of Strain

In addition to the mechanisms we have already discussed (proximity effects, orientation effects, general acid-base catalysis, and nucleophilic catalysis), one further factor that has been widely cited as a possible contributor to specific rate enhancements in enzyme catalysis is the induction of strain in the substrate, the enzyme, or both when the enzyme-substrate complex is formed. The arguments have been stated in many ways,[1] but for our purposes the following simplified picture should suffice.

It is certainly true in general that the substrate in an enzyme-catalyzed reaction (and perhaps the enzyme as well) undergoes significant structural alterations as it is transformed through a transition state (or several transition states with intervening metastable intermediates) to the product(s) of the reaction while bound at the enzyme active site. It is also reasonable to assume that the binding forces between the enzyme active site and the reacting substrate will vary as a function of these structural changes as the reaction proceeds. At some point in the conversion of substrate to product, these binding forces will be maximal. It is entirely possible that this optimal binding might occur when the substrate reaches the transition state for the rate-determining step of the reaction. If so, this in itself could provide a powerful driving force for the reaction. In effect, the enzyme in such a case would be using potential binding energy to help force the substrate toward the critical transition state structure. In other words, the enzyme upon binding the substrate would induce strain on the substrate structure in the direction of the transition state structure, thus facilitating the reaction.

The exact nature of these strain forces is not well defined. It seems likely, however, that they involve noncovalent interactions between enzyme and substrate such as the hydrophobic and hydrogen bonding interactions discussed in Chapter 1. An additional type of strain force, termed "electronic distortion," has been discussed in a series of papers by Zeffren and co-workers,[35-37] who suggested that the influence of ion pairs and dipoles at an enzyme's binding site could be to decrease the gap between the energy levels of the substrate's ground state and the transition state for the enzymatic reactions. Since binding sites are typically considered to be substantially nonaqueous, ion pairs and dipoles present there would not be hydrated and hence their effects would be much stronger than would normally be the case in an aqueous environment.[36]

Looking at it in a slighty different way, the function of the enzyme might be to use its binding forces to stabilize the transition state more effectively than it stabilizes the substrate or product. This means that the energy gap between the enzyme-substrate complex ES and the enzyme-transition state complex ES‡ (i.e., the free energy of activation for the reaction) will necessarily be smaller than the corresponding energy gap between S and S‡ in the uncatalyzed reaction. Hence we have a rate enhancement.

One of the most convincing types of evidence that the kind of phenomenon just described may in fact occur is the observation of remarkably effective competitive inhibition of certain enzymes by molecules which are structural analogs not of substrates but rather of presumed *transition states* (or presumed reactive intermediates which should resemble transition states more closely than substrates or products). Consider the enzyme cytidine deaminase from *E. coli*. It catalyzes the interconversion of cytidine (**I**) and uridine (**III**), and the reaction is presumed to involve a nucleophilic displacement through a tetrahedral intermediate such as **II**. It has been found[38] that 3,4,5,6-tetrahydrouridine (**IV**), a stable compound which, like the presumed intermediate (**II**), has a tetrahedral carbon in position-4 of the pyrimidine ring, is an exceedingly potent competitive inhibitor of the enzyme. In fact its competitive inhibition constant K_i is some 3000 times smaller than that of uridine (**III**), 6000 times smaller than that of 5,6-dihydrouridine (**V**), and over 10,000 times smaller than K_m for cytidine itself. No known substrate-analog for cytidine deaminase even approaches this effectiveness as an inhibitor.

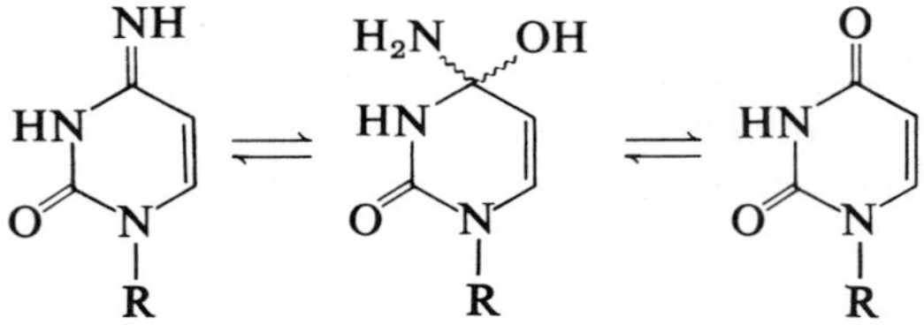

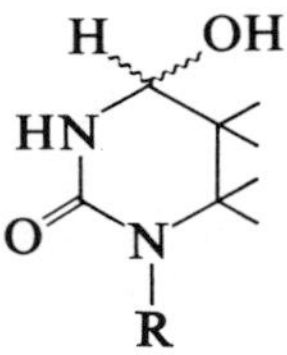

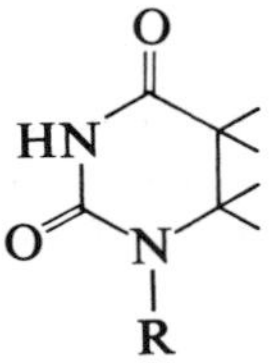

Another dramatic example of inhibition by a transition state analog is the potent inhibition of triosephosphate isomerase by 2-phosphoglycolate (**VI**).[38,39] The enzyme catalyzes the interconversion of triosephosphates **VII** and **IX**, presumably through a high-energy ene-diolate intermediate indicated by **VIII**.[40] In this case, the putative transition state analog **VI**

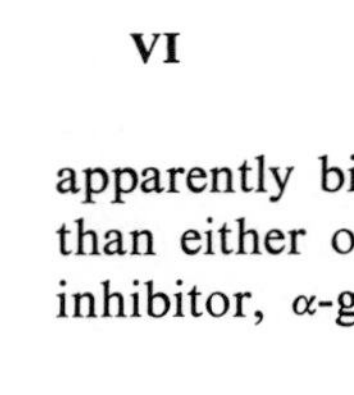

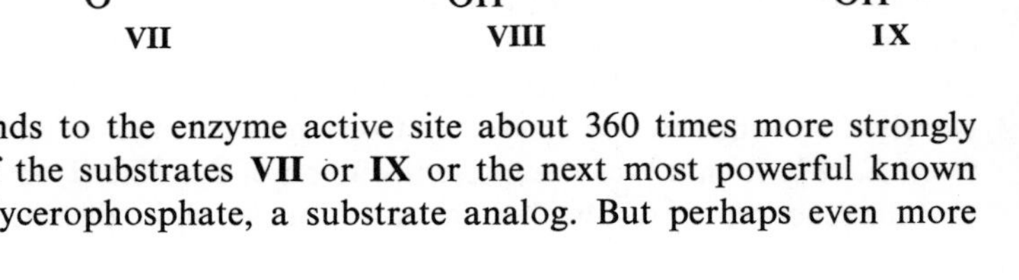

apparently binds to the enzyme active site about 360 times more strongly than either of the substrates **VII** or **IX** or the next most powerful known inhibitor, α-glycerophosphate, a substrate analog. But perhaps even more significantly, the pH dependence of K_i for **VI** parallels not the pH dependence of K_m for **VII** but rather the pH dependence of the *catalytic* parameter k_{cat}. This strongly indicates that the binding of transition state analog **VI** is sensitive to the state of ionization of an enzyme residue which appears to be specifically involved in *catalysis* as opposed to initial substrate binding (see Chapter 5).

References

1. W. P. Jencks, *Catalysis in Chemistry and Enzymology*, McGraw-Hill, New York, 1969.
2. D. E. Koshland, Jr., *J. Theoret. Biol.*, **2**, 75 (1962).
3. D. E. Koshland, Jr., and K. E. Meet, *Ann. Rev. Biochem.*, **27**, 359 (1968).
4. T. C. Bruice and S. J. Benkovic, in *Bioorganic Mechanisms*, Vol. 1, Benjamin, New York, 1966, Chap. 1.
5. T. C. Bruice, in *The Enzymes*, Vol. II, P. D. Boyer (Ed.), Academic, New York, 1970, Chap. 4.
6. T. C. Bruice and A. Turner, *J. Am. Chem. Soc.*, **92**, 3422 (1970).
7. H. R. Mahler and E. H. Cordes, *Biological Chemistry*, Harper and Row, New York, 1966, p. 299.
8. T. C. Bruice and S. J. Benkovic, *J. Am. Chem. Soc.*, **85**, 1 (1963).
9. T. C. Bruice and A. Turner, *J. Am. Chem. Soc.*, **92**, 3422 (1970).
10. D. R. Storm and D. E. Koshland, Jr., *Proc. Nat. Acad. Sci., U.S.*, **66**, 445 1970.
11. T. C. Bruice, A. Brown, and D. O. Harris, *Proc. Nat. Acad. Sci. U.S.*, **68**, 658 (1971).
12. J. Reuben, *Proc. Nat. Acad. Sci. U.S.*, **68**, 563 (1971).
13. Ref. 1, pp. 44–67.
14. See ref. 1, Chap. 10 and ref. 4 for detailed discussions.
15. T. C. Bruice and G. L. Schmir, *Arch. Biochem. Biophys.*, **63**, 484 (1956).
16. B. M. Anderson, E. H. Cordes, and W. P. Jencks, *J. Biol. Chem.*, **236**, 455 (1961).
17. G. DiSabato and W. P. Jencks, *J. Am. Chem. Soc.*, **83**, 4393 (1961).
18. T. C. Bruice and S. J. Benkovic, *Bioorganic Mechanisms*, Vol. II, Benjamin, New York, 1966, pp. 35–59.
19. E. S. Gould, *Mechanism and Structure in Organic Chemistry*, Holt, Rinehart and Winston, New York, 1959, p. 346; Ref. 1, p. 116.
20. J. P. Guthrie and F. H. Westheimer, *Fed. Proc.*, **26**, 562 (1967).
21. S. Warren, B. Zerner, and F. H. Westheimer, *Biochemistry*, **5**, 817 (1966).
22. F. H. Westheimer and H. Cohen, *J. Am. Chem. Soc.*, **60**, 90 (1938); T. A. Spencer, H. S. Neel, T. W. Flechtner, and R. A. Zayle, *Tetrahedron Lett.*, **1965**, 3889.
23. B. L. Horecker, S. Pontremoli, C. Ricci, and T. Cheng, *Proc. Nat. Acad. Sci. U.S.*, **47**, 1949 (1961).

24. J. C. Speck, Jr., P. T. Rowley, and B. L. Horecker, *J. Am. Chem. Soc.*, **85**, 1012 (1963).
25. E. H. Cordes and W. P. Jencks, *J. Am. Chem. Soc.*, **84**, 826 (1962).
26. E. H. Cordes and W. P. Jencks, *Biochemistry*, **1**, 773 (1962).
27. Ref. 1, pp. 173–182.
28. L. P. Hammett, *Physical Organic Chemistry*, 2nd ed., McGraw-Hill, New York, 1970, Chaps. 10 and 11.
29. Ref. 1, pp. 182–199.
30. E. H. Cordes and W. P. Jencks, *J. Am. Chem. Soc.*, **85**, 2843 (1963).
31. C. G. Swain and J. F. Brown, Jr., *J. Am. Chem. Soc.*, **74**, 2534, 2538 (1952).
32. S. Winstein, S. Smith, and D. Darwish, *J. Am. Chem. Soc.*, **81**, 5511 (1959).
33. A. M. Eastham, E. L. Blackall, and G. A. Latremouille, *J. Am. Chem. Soc.*, **77**, 2182 (1955); E. L. Blackall and A. M. Eastham, *J. Am. Chem. Soc.*, **77**, 2184 (1955); Y. Pocker, *Chem. Ind. (London)*, **1960**, 968.
34. M. F. Perutz, *Eur. J. Biochem.*, **8**, 455 (1969).
35. E. Zeffren and R. E. Reavill, *Biochem. Biophys. Res. Commun.*, **32**, 73 (1968).
36. E. Zeffren, *Arch. Biochem. Biophys.*, **137**, 291 (1970).
37. C. E. Stauffer and E. Zeffren, *J. Biol. Chem.*, **245**, 3282 (1970).
38. R. Wolfenden, *Nature (London)*, **223**, 740 (1969).
39. R. Wolfenden, *Biochemistry*, **9**, 3404 (1970).
40. I. A. Rose, *Brookhaven Symposia in Biology*, **15**, 293 (1962).

CHAPTER 8

Organic Coenzymes

As pointed out in Chapter 1, many enzyme reactions require the presence of an additional nonproteinaceous compound in order for catalysis to occur. This material can either be a metal ion, as, for example, the essential zinc(II) atom of carboxypeptidase A or carbonic anhydrase, or it can be an organic material bound more or less tightly to the enzyme protein. Indeed, many enzymes require both a metal ion and an organic "cofactor(s)." This chapter is concerned with the organic cofactors most commonly encountered in enzymology. Before proceeding further, however, let us consider cofactor terminology to be used here and coenzyme function in biochemistry.

a. Cofactor Terminology

There is some ambiguity in biochemical literature concerning the definitions of the terms cofactor, coenzymes, prosthetic groups, carrier, and substrate. Basically, any substance required for the manifestation of enzymic activity which emerges from the reaction unchanged may be considered a *cofactor* for that enzyme. For example, the nontransition metal ions or inorganic phosphate have frequently been found necessary for catalysis to occur in a given enzyme system; hence these are both cofactors. Contrary to the coenzymes described below, however, these materials do not appear to be involved in the catalytic events of enzyme function. Rather, their role seems to be one of maintaining the enzyme in an active configuration. In this book, the term coenzyme is reserved for those cofactors of basically an organic chemical nature, such as pyridoxal or nicotinamide adenine dinucleotide (NAD$^{\oplus}$).

Organic coenzymes are generally found to undergo some sort of reversible chemical transformation during the catalytic process in much the same way as the enzymes themselves presumably do, and what distinguishes a coenzyme from a substrate is the eventual regeneration of the original form of the coenzyme. This regeneration may occur while the coenzyme is firmly bound to its original enzyme protein; in this case the enzyme is referred to as a *prosthetic group*. Pyridoxal phosphate is an example of such a coenzyme. In other cases, the coenzyme undergoes some sort of transformation, for example, reduction, while bound to one enzyme. Then it dissociates from this enzyme as the reduced form, and migrates to another enzyme where it also serves as a cofactor and is changed back to its original form (oxidized) during this second process. In this case, the coenzyme is said to be a *carrier molecule* and in the reduction/oxidation example the species carried is a "reducing equivalent"—an electron. One can see, then, that a series of reduction/oxidation steps like this could and does form the basis of overall electron transport in biological systems (see Chapter 10). NAD$^\oplus$ is an important example of a carrier coenzyme.

Thus the chief difference between a prosthetic group and a carrier is the site of regeneration of the original coenzyme structure.[1] For prosthetic groups, this site is the same enzyme upon which it first reacted; for carrier groups, it is on a different enzyme. A substrate is distinct from both of these in that it generally is not immediately regenerated in a subsequent enzymic step by the enzyme system being studied; rather, it usually undergoes further change.

b. Coenzyme Function

As is apparent from studying coenzyme structure, many coenzymes either are themselve nucleotides (ATP, UTP, etc.) or bear strong resemblance to nucleotides (FMN, NAD$^\oplus$) (see Table 8-1). One could speculate on this basis that a coenzyme of this type might serve as a partial template for holding the enzyme protein in a catalytically active state and is bound tightly because of its resemblance to the nucleic acid template upon which the protein's synthesis was directed. This may indeed be the origin of tight binding. However, such binding to maintain a protein conformation cannot be the sole function of a coenzyme since it does not explain the chemical transformations that occur on the coenzyme molecule.

What purposes, then, does a coenzyme serve? If one studies Table 8-1 it may be seen that coenzymes serve one of two vitally important purposes. Either they are involved in *energy release (1–6) and transfer (7)* or they are crucially involved in *biosynthetic processes (8–14)*.

In general, all reactions involving coenzymes utilize the coenzyme to *transfer a chemical entity* from one molecule to another. In the enzyme systems that release energy, the species transferred is an electron pair; since protons usually accompany the electrons (either as part of a hydride transfer or as abstracted from the aqueous surroundings), this process is called either electron transfer or hydrogen transfer. The nucleoside di- and triphosphates are the coenzymes involved in energy transfer. These crucial coenzymes are involved in all the transphosphorylation reactions, which allow the utilization of the chemical energy of the polyphosphate linkages. The other coenzymes of Table 8-1 generally are involved in the transfer of one carbon or two carbon atom fragments in a biosynthetic pathway. Thus we see biotin transferring the carboxyl group and coenzyme A transferring (primarily) the acetyl group. Occasionally, larger groups are transferred—ATP is known in some reactions to transfer its adenosine group; usually, however, the coenzymes of biosynthesis transfer smaller fragments.

c. The Relationship between Coenzymes and Vitamins

The term "vitamin" has been omitted from our discussion thus far. It has long been recognized that certain dietary deficiencies lead to certain diseases characteristic of the deficiency: dietary niacin (nicotinamide) deficiency leads to pellagra, ascorbate deficiency leads to scurvy, and so on. Vitamins, then, are substances we must obtain in our diet. We do not have the biochemical machinery to synthesize them, and without them a diseased state occurs. It has been found that a great many vitamins are coenzymes themselves or form an essential part of a coenzyme. For example, vitamin C or ascorbic acid, itself is a coenzyme involved in hydroxylation of proline and lysine residues in collagen.[2] Nicotinamide, a B vitamin, is the functional portion of $NAD^{\oplus}$. Folic acid, also a B vitamin, must be supplied in the diet in order for the body to synthesize the coenzyme tetrahydrofolic acid.

It is tempting to generalize and say that all vitamins are themslves coenzymes or coenzyme precursors. At this point such a generalization cannot be made since there are vitamins with no presently known coenzyme function. For example, vitamin E, α-tocopherol (I), is presently thought to be an

I

antioxidant, or radical scavenger to prevent lipid peroxidation.[3] It is interesting to note, however, that until the mid-1960s, vitamin C was also thought to function mainly as a biological antioxidant.[1] As mentioned earlier, it is now known to have a distinct coenzymatic function.[2] It will require much diligent effort to determine if all vitamins serve some coenzyme function; clearly, very many are known to do so presently.

The Individual Coenzymes

The remainder of this chapter presents in tabular form the major organic coenzymes. The coenzymes are listed according to reaction type [electron,

TABLE 8-1 The Common Organic Coenzymes

No.	Name	Structure	Transferred Groups		Classification
			Group	Site Involved	
1a	Nicotinamide adenine dinucleotide (NAD^+) (reduced form: NADH)		$H^{\oplus}, 2e^{\ominus}$	C4-Nicotinamide ring	Carrier
1b	Nicotinamide adenine dinucleotide Phosphate ($NADP^+$) (Reduced form: NADPH)	Same as NAD^+ + additional phosphate group at C2 of adenine ribose.	$H^{\oplus}, 2e^{\ominus}$	C4-Nicotinamide ring	Carrier
2	Ascorbic acid (Vitamin C)		$2H^{\oplus}, 2e^{\ominus}$	Not well understood; probably C2 and C3	Carrier

3	Quinones[a]: ubiquinone (coenzyme Q)	$2H^{\oplus}, 2e^{\ominus}$	Quinone carbon atoms	Carrier
4	Gluthathione (GSH) (oxidized form: GSSG)	$H^{\oplus}, 2e^{\ominus}$	–SH	Carrier
5	Heme coenzymes:[b] (a) Cytochrome systems	$H^{\oplus}, e^{\ominus}$	Fe(III)-Heme	Carrier
	(b) Hydroperoxidases: (i) Peroxidase (ii) Catalase	$2H^{\oplus}, 2e^{\ominus}$	Fe(III)-Heme	Prosthetic
6a	Flavin mononucleotide (FMN)	$H^{\oplus}, 2e^{\ominus}$	Amidine C atom, C‡	Prosthetic

(continued)

TABLE 8-1 (*continued*)

No.	Name	Structure	Transferred Groups		Classifi- cation
			Group	Site Involved	
6b	Flavin adenine di-nucleotide (FAD)		H$^\oplus$, 2e$^\ominus$	Amidine C atom, C‡	Prosthetic
7	Nucleoside Di- and Tri-phosphates^e (ADP, UDP, CDP, IDP, GTP); (ATP, UTP, CTP, ITP, GTP)	(ATP)	Pi, PPi, others^e		Carrier

8	Thiamine pyro-phosphate (TPP)		$RC\underset{O}{\overset{\parallel}{C}}$ (usually $R=CH_3$)	C_2 Thiazolium ring	Prosthetic
9	Lipoic acid		$RC\underset{O}{\overset{\parallel}{C}} + 2e^{\ominus}$	SH	Carrier
10	Coenzyme A (CoASH) (acyl-ated form): $RC\underset{O}{\overset{\parallel}{C}}-SCoA$		$RC\underset{O}{\overset{\parallel}{C}}$ (usually $R=CH_3$)	SH	Carrier
11	Biotin[d]		CO_2	N_1 or ureido O	Carrier[d]

(continued)

TABLE 8-1 (continued)

No.	Name	Structure	Transferred Groups		Classifi-cation
			Group	Site Involved	
12	Tetrahydrofolic acid (FH₄; THFA)	(structure)	—CH₃ —CH₂OH —CHO —CH=NH	N5 N5–N10 N5–N10; N10, N5 N5–N10; N10, N5	Carrier
13	Cobalamin (vitamin B₁₂)	Very complex (see text)	H⊖	Co(III) + C5 of 5-deoxyadenosyl moiety	Carrier
14	Pyridoxal phosphate[e] (PLP)	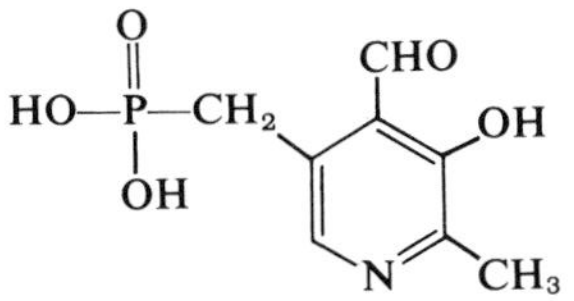 (structure)	—NH₂; O[e]	Aldehyde group	Prosthetic

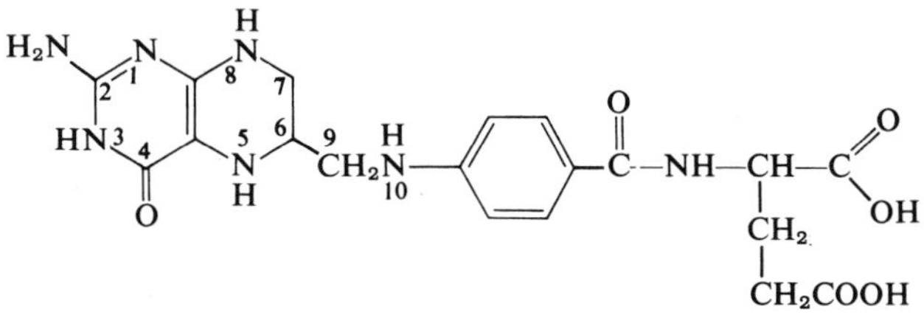

[a] Vitamins K are naphthoquinone derivatives.

[b] Nonprotein portion of heme coenzymes is the heme group for both cytochromes and hydroperoxidases. Heme-protein complexes within the cytochrome system serve as the total carrier. Enzymes involved are the cytochrome oxidases. Nature of heme group substituents varies considerably. Structure shown is ferric protoporphyrin IX, the prosthetic group of the hydroperoxidases.

[c] A = Adenine, U = Uridine, C = Cytidine, I = Inosine, G = Guanosine. ATP is structure shown. These coenzymes transfer a wide variety of groups, depending on site of cleavage and nucleotide under discussion. See text. See also reference 4.

[d] Site of biotin carboxylation in vivo not conclusively established; see text. Recent work has shown that biotin is probably attached covalently to a small protein through its pentanoyl carboxyl group. This "carboxyl carrier protein" is thus seemingly analogous to the cytochrome system; hence its classification as a carrier.

[e] Oxygen transfer noted here is a formal representation of the group transferred during transamination of α-amino acids to α-keto acids.

transfer (hydrogen transfer), phosphate transfer, carbon fragment transfer, and amino and oxo transfer] and classified as to whether they are carriers or prosthetic groups. Discussion of the individual coenzymes will be brief except for $NAD^{\oplus}$ systems and pyridoxal phosphate systems. These will be treated in greater detail to exemplify coenzyme studies.

Coenzymes for Energy Release and Transport

a. Nicotinamide Adenine Dinucleotide (NAD⊕)

One of the clearest ways to demonstrate the necessity for some sort of cofactor in an enzymic system is to find a set of experimental conditions that causes the cofactor to dissociate from its enzyme and then dialyze the enzyme under these conditions. If this can be done, then (1) the dialyzed enzyme shall have no activity under normal assay conditions, (2) the dialyzate should have no activity, and (3) addition of dialyzate to enzyme should result in the regeneration of activity. In this manner, $NAD^{\oplus}$ was recognized in 1904 as a necessary dialyzable cofactor required for the conversion of glucose to ethanol by cell-free extracts of yeast. This was the first coenzyme to be recognized. As $NAD^{\oplus}$ or $NADP^{\oplus}$, these extremely important coenzymes of electron transport are involved in a large number of dehydrogenation reactions, during which process the reduced form of the coenzyme [NAD(P)H] is formed. For example, the alcohol dehydrogenases catalyze the reaction of Eq. 8-1. It is usually observed that enzymes utilizing $NAD^{\oplus}$ coenzymes are

$$RCH_2OH + NAD^{\oplus} \rightleftharpoons RCHO + NADH + H^{\oplus} \qquad (8\text{-}1)$$

specific for one form or the other, that is, they either utilize $NAD^{\oplus}$ or $NADP^{\oplus}$, but not both.

Among the important questions to be asked when studying a coenzyme system is: What is the site involved in the catalytic function and what processes occur there? Isolation and identification of NADH as the product of a dehydrogenation reaction has made it clear that the process that occurs with $NAD^{\oplus}$ or $NADP^{\oplus}$ is the reduction of the nicotinamide ring of the coenzyme; it is not immediately obvious, however, which of the three possible products, **IIa**, **IIb**, or **IIc**, is formed. Colowick and co-workers[5] have provided an

IIa (C2)   **IIb (C4)**   **IIc (C6)**

answer to this question with a combination of chemical and enzymic experiments. First they reduced NAD$^{\oplus}$ in D_2O to form deuterated reduced NADD (**III**). This was then converted *enzymatically* to a deuterium-labeled NAD$^{\oplus}$

(**IV**). This was then chemically converted to the deuterated N-methylnicotinamide derivative **V**. It had been known that **V** is oxidized predominantly at the 2 and 6 positions of the nicotinamide ring, hence the appearance of D in the product pyridone derivatives showed that the site of reduction in NAD$^{\oplus}$ systems is apparently C4. The word "apparently" is used here because Chaykin[6] has pointed out that (1) the position α to the pyridine N atom is the kinetically favored site of anion (including H$^{\ominus}$) addition, (2) electron-withdrawing substituents at C3 (such as carboxamido) favor attack at C2, and (3) 1,2 or 1,6 addition products will isomerize to the thermodynamically more stable 1,4 adducts. Thus the formation of a 1,2 adduct cannot be completely excluded. It should be noted, however, that if it does form, it only has a fleeting existence and must rapidly isomerize in a stereospecific manner to the C4 derivative. This is so because 1,2 adducts (λ_{max} 395 nm) are spectrally distinct species from 1,4 adducts (λ_{max} 345 nm) and only 1,4 adducts have been observed in steady-state studies. The use of rapid reaction techniques would be useful in this regard.*

---

* Hamilton[29] has proposed an intriguing mechanism for dehydrogenations requiring NAD$^{\oplus}$. In this proposal the substrate AH$_2$, acting as a nucleophile, attacks the C2 position of the nicotinamide ring to form an intermediate which can then transfer a proton (or hydride ion) stereospecifically to C4, as shown in Eq. i. As drawn, both hydrogens are transferred as protons, although in the cyclic intermediate (1), it is impossible to distinguish between H-transfer as shown and H-transfer as hydride ion (arrows reversed). For theoretical reasons, proton transfer is preferred. (See Chapter 10.)

The isomerization mentioned earlier, if indeed it is needed, must occur in a stereospecific manner. This is required in view of the fact that all dehydrogenations of NADH and hydrogenations of $NAD^{\oplus}$ are stereospecific. Vennesland and Westheimer[7] demonstrated that dehydrogenases stereospecifically add deuterium to $NAD^{\oplus}$ to produce what is termed either the A- or the B-NADH structure but never (with one known exception) both. These structures are shown in **VIa** and **VIb**.

VIa   A-NADH   **VIb   B-NADH**

The fact that either the A or B derivatives are formed indicates that each enzyme interacts with $NAD^{\oplus}$ in its own characteristic way. When considering these interactions the structure of $NAD^{\oplus}$ or $NADP^{\oplus}$ free in solution also has to be considered. Fluorescence studies have shown that there is an interaction between the adenine and nicotinamide rings of these molecules. Thus excitation of the adenine ring led to the observation of the emission spectrum of nicotinamide.[8] This behavior is lost at higher temperature, implying that at low temperatures a stacked ring conformation exists, and at higher temperatures an open conformation exists. NMR studies by Kaplan[8] have confirmed this and have also given evidence of restricted rotation about the nicotinamide-ribose linkage.

Kaplan's penetrating NMR studies have also shown that the oxidized and reduced forms of $NAD^{\oplus}$ have different conformations in solution. This probably explains the great binding preferences of dehydrogenases for $NAD^{\oplus}$ and hydrogenases for NADH. Indeed, the A and B forms of NADH can be distinguished via high-resolution NMR studies.[8] These studies have led Kaplan to propose that *in vivo*, three forms of $NAD^{\oplus}$ exist. Two of these,

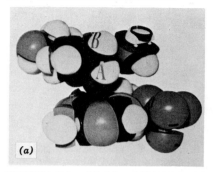 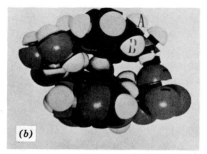

(a) *(b)*

FIGURE 8-1. Models of NADPH. *(a)* The *P* helix of NADPH. *(b)* the *M* helix of NADPH. (Source: Kaplan and Sarma, ref. 8.)

called the *P*-helical form and *M*-helical forms, are thought to be active co-enzymatically, and they are thought to be related to each other by equilibrium through a nonactive, open form. This is shown schematically in Figure 8-2. Figure 8-1 shows molecular models of the *P*- and *M*-helical forms of NADH. These figures show clearly why an enzyme which binds only one helical form would be able to distinguish between the two protons of the coenzyme (or why, for NAD$^{\oplus}$, stereospecific addition of hydrogen would occur). It is conceivable that the enzyme that adds to either side of NAD$^{\oplus}$ (lipoyl dehydrogenase) binds the open form of the coenzyme.

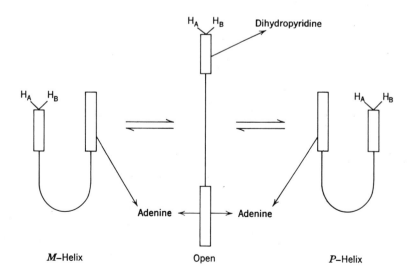

FIGURE 8-2. Equilibrium of the different forms of NADH.

One structural point of importance in NAD⊕ binding is the influence of the carboxamido group. NMR studies on NAD⊕ analogs where this group has been replaced by either a formyl, butyl, or amino group have shown that the overall geometry of the NAD⊕ molecule is the same for all these derivatives; yet they vary markedly in their abilities to act as coenzymes. This indicates that NAD⊕ dependent dehydrogenases can finely distinguish what is present at the C3 site of the nicotinamide ring.

b. Ascorbic Acid

This coenzyme is the well-known vitamin C present in citrus fruits and necessary for the prevention of scurvy in man. It was first crystallized by Szent-Gyorgi in 1928.[4] Only the L-form is active and the product of its oxidation in biological tissues is termed L-dehydroascorbic acid (VII). It has recently been recognized to have a coenzymatic function which is directly involved

VII

in the prevention of scurvy. This function allows the enzyme called collagen proline hydroxylase to form hydroxyproline from collagen-bound proline (not from free proline).[9] It seems likely that ascorbic acid will also be found to be functional in the hydroxylation of lysine in collagen as well. Without these vitamin C-mediated reactions, collagen synthesis is faulty, and connective tissue lesions as seen in scurvy are the result.[10]

Although it is not clear what enzymatic reactions are involved, ascorbic acid has also been suggested to aid in the prevention of atherosclerosis.[10][12] Similarly, its possible role in the prevention of the common cold is unclear. In this last area, ascorbic acid has received quite a bit of notoriety;[13,14] definitive studies in this area are lacking.

c. Quinones

These materials as exemplified by the structure in Table 8-1 play an important role in the respiratory system as intermediate electron carriers. There are two classes—the ubiquinones, where the quinoidal ring is methoxylated as shown in the table, and the vitamin K series, as shown in VIII. Ubiquinone itself is the structure in Table 8-1 where $n = 10$, and is also known

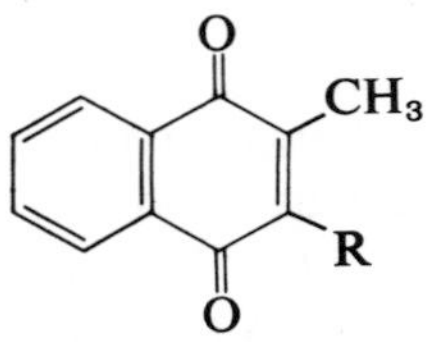

Menadione R = H

Vitamin K_1 R = $CH_2CH{=}CCH_2(CH_2CH_2CHCH_2)_3H$

Vitamin K_2 R = $(CH_2CH{=}CCH_2)_nH$

VIII

as coenzyme Q. During enzymatic function the quinones are reduced to hydroquinones and regenerated by a system involving cytochrome C and cytochrome oxidase (see later).

d. Glutathione (GSH)

The most widely studied coenzymatic function of GSH is in the glyoxylase reaction, the hydrogenation of methylglyoxal to lactate. This has been shown to proceed by the pathway of Eq. 8-2. Support for this mechanism comes from

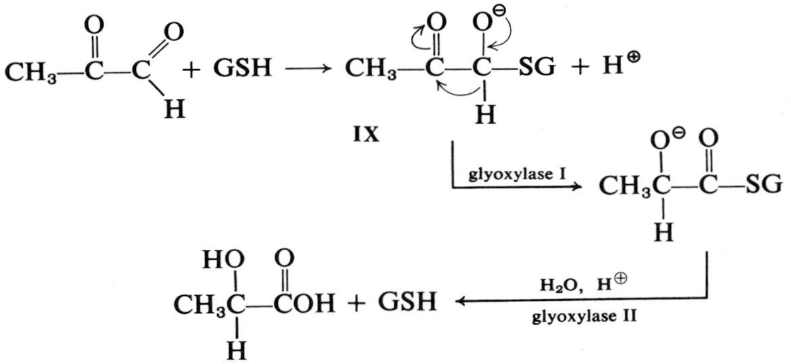

$$CH_3{-}\overset{O}{\underset{}{C}}{-}\overset{O}{\underset{H}{C}} + GSH \longrightarrow CH_3{-}\overset{O}{\underset{}{C}}{-}\overset{O^{\ominus}}{\underset{H}{C}}{-}SG + H^{\oplus}$$

IX

$$\xrightarrow{\text{glyoxylase I}} CH_3\overset{O^{\ominus}}{\underset{}{C}}{-}\overset{O}{\underset{H}{C}}{-}SG \qquad (8\text{-}2)$$

$$CH_3\overset{HO}{\underset{H}{C}}{-}\overset{O}{\underset{}{C}}OH + GSH \xleftarrow[\text{glyoxylase II}]{H_2O,\ H^{\oplus}}$$

the following observations: (1) the protonated form of **IX** is a substrate for glyoxylase I; (2) the H atom at C2 of lactate has been shown to come from Cl.

In a possibly nonenzymic way, GSH is thought to act also as an oxygen scavenger, as in Eq. 8-3, which would allow sulfhydryl functional enzymes to remain in their reduced state. This protective function has not been conclusively demonstrated.

$$4GSH + O_2 \longrightarrow 2GSSG + 2H_2O \qquad (8\text{-}3)$$

e. Heme Coenzymes

Structure **X** is a general representation of the porphyrin ring of the heme coenzymes. The ferrous derivatives are called *heme* groups, the ferric derivatives are called *hematin* groups. Although other metals are found in some

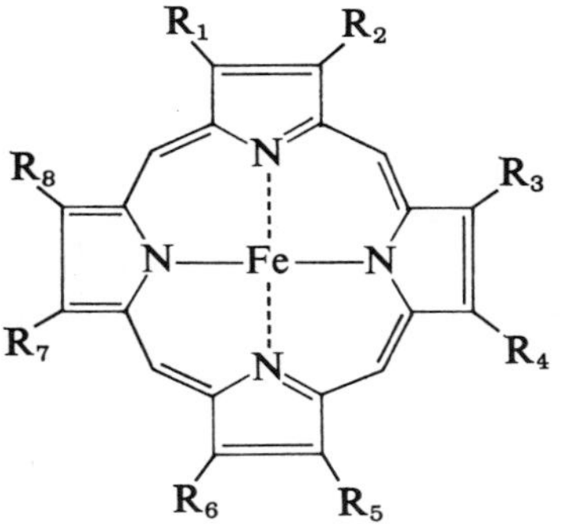

heme = ferrous
hematin = ferric

X

systems, iron is clearly the predominantly utilized metal ion in mammalian systems. The structure shown in the table is that for protoporphyrin **IX**, where R_1, R_3, R_5, and $R_8 = CH_3$; R_2, $R_4 = $ —CH=CH_2, and R_6 and $R_7 = CH_2CH_2COOH$. This important derivative is the heme structure for the oxygen-carrying proteins hemoglobin and myoglobin and for the hydroperoxidases catalase and peroxidase.

In the cytochromes, which are proteins involved in respiratory electron transport to flavoproteins, R may be hydrogen or C_2H_5 as well as those substituents in protoporphyrin **IX**. The transfer of electrons from the cytochromes is mediated by enzymes called cytochrome oxidases. Regeneration occurs with another enzyme—the cytochrome reductases. Thus this is reminiscent of the electron transfer function of $NAD^\oplus$. In this case, however, the carrier is not just a complex organic species, it is also tightly bound to its "own" protein.

In the hydroperoxidases, Blumberg and co-workers[15] have shown by spectral means that the ferric ion [Fe(III)] is first oxidized to Fe(V) (Compound I) during the first reaction with H_2O_2; this is then reduced in two one-

$$\text{Peroxidase Fe(III)} + H_2O_2 \longrightarrow [\text{Peroxidase-Fe(V)}\cdot\text{OOH}] + H^\oplus$$

$$\underset{\text{Compound I}}{} \qquad\qquad\qquad AH_2$$

$$\begin{array}{c}\text{Peroxidase Fe(III)} + A \longleftarrow \text{Peroxidase-Fe(IV)}\cdot\text{OH}\cdot\text{AH} \\ + \qquad\qquad\qquad\qquad + \\ H_2O \qquad\qquad\qquad\qquad OH^\ominus\end{array} \qquad (8\text{-}4)$$

electron reduction steps by the substrate AH_2 (Eq. 8-4). This does not require any other protein, making the hematin a true prosthetic group.

Thus heme coenzymes are involved in three crucial processes, oxygen transport and utilization, electron transport, and peroxide cleavage.

f. Flavins (FMN, FAD)

The key role of the flavoproteins in many life processes is the transfer of reducing equivalents in a great many dehydrogenation reactions. The oxidation of many amino acids, sugars, and intermediates in fatty acid and carbohydrate metabolism requires a flavoenzyme. In addition, flavoproteins also serve as electron acceptors for many reduced coenzymes in order to allow the regeneration of the oxidized form of the coenzyme (e.g., lipoamide dehydrogenase, quinone reductases, NADH rectuctases). For example, the enzyme "Reduced NADP$^\oplus$ dehydrogenase" (EC 1.6.99.1) is a flavin mononucleotide requiring enzyme for which the first step is the reduction of FMN to FMNH and the associated oxidation of NADPH to NADP$^\oplus$. Cytochrome c serves as an acceptor of the reducing equivalents, generating reduced cytochrome c and the regenerated flavoenzyme. When a specific substrate (e.g., an amino acid, but not a coenzyme) provides the reducing equivalents (i.e., it is oxidized), and when oxygen is reduced in the regeneration of the original flavin, one has an oxidase. Thus Eq. 8-5 is drawn for the amino acid oxidases (EC 1.4.3.2, 1.4.3.3). The chemical mechanisms of the oxidases are discussed further in Chapter 10. (See also Ref. 29).

$$
\begin{array}{c}
\text{NH}_2 \\
| \\
\text{R—CH—COOH} + \text{O}_2 + \text{Enzyme·FAD} \longrightarrow
\end{array}
$$

$$
\begin{array}{c}
\text{O} \\
\| \\
\text{R—C—COOH} + \text{Enzyme·FADH}_2·\text{O}_2 \\
\\
\text{H}_2\text{O}_2 + \text{Enzyme·FAD} \longleftarrow
\end{array}
$$

(8-5)

A new role for the flavins is just being recognized. This in the involvement of flavoproteins as replacements for the ferredoxins in electron transport. Such species have been termed "flavodoxins."[16]

g. Nucleotide Coenzymes

The first six coenzymes of Table 8-1 are involved in energy release through electron (H atom) transfer processes. The enzymes that utilize these coenzymes are briefly discussed from a mechanistic viewpoint in Chapter 10; such enzymes typically require a transition metal ion as well as a coenzyme. The nucleotide di- and triphosphates are involved primarily with the transfer of energy through the synthesis and cleavage of polyphosphate linkages. The free energy derived from the hydrolysis of these phosphoanhydride linkages ultimately becomes available to drive to completion an otherwise unfavorable process. During the process of a reaction involving, say, ATP, a portion of the ATP molecule is transferred to a substrate, leaving some remainder of the ATP. Thus in the synthesis of adenosine, ADP is formed when ATP transfers an inorganic phosphate group to ribose (Eq. 8-6). The advantage to the host organism in utilizing Eq. 8-6 to synthesize adenosine,

$$\text{Ribose} + \text{ATP} \xrightarrow[E_1]{Mg^{2+}} \text{Ribose-1-phosphate} \xrightarrow[E_2]{\text{Adenine}} \text{Adenosine} + \text{Pi} \quad (8\text{-}6)$$
$$+$$
$$\text{ADP}$$

rather than to do it directly from ribose and adenine, is that the phosphate at Cl of ribose is a much better leaving group than is hydroxide in a displacement reaction such as this. Hence energy has been transferred from ATP to ribose, leading to an overall more favorable process for nucleoside synthesis.

Structure **XI** shows the different types of transfer reactions known to occur with a nucleoside triphosphate. The example shown is adenosine triphosphate. The reader is directed to reference 4 for a broader discussion of these coenzymes.

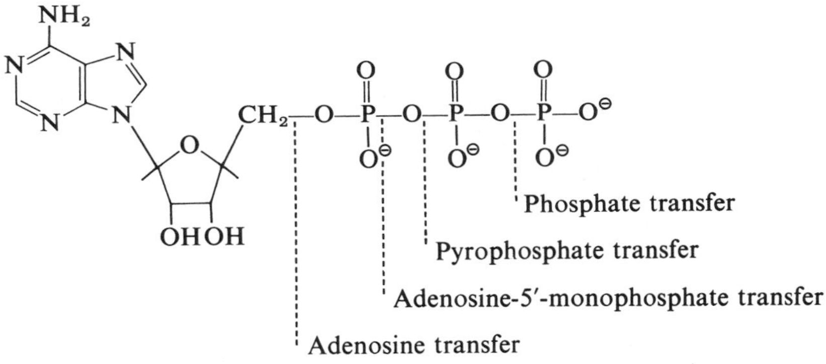

Phosphate transfer

Pyrophosphate transfer

Adenosine-5′-monophosphate transfer

Adenosine transfer

XI

Coenzymes for Biosynthesis

a. Thiamine Pyrophosphate (TPP)

This coenzyme is primarily involved in three types of reactions: (1) oxidative decarboxylations of α-keto acids; (2) nonoxidative decarboxylation of α-keto acids; and (3) the formation of α-keto alcohols. Examples are shown in Eq. 8-7.

Oxidative Decarboxylation

$$\underset{\text{R}-\overset{\overset{\textstyle O}{\|}}{\text{C}}-\text{COOH}}{} + \tfrac{1}{2}O_2 \longrightarrow \underset{\text{R}\overset{\overset{\textstyle O}{\|}}{\text{C}}-\text{OH}}{} + CO_2 \qquad (8\text{-}7a)$$

Nonoxidative Decarboxylation

$$R—\overset{\overset{\displaystyle O}{\|}}{C}—COOH \longrightarrow R\overset{\overset{\displaystyle O}{\|}}{C}—H + CO_2 \tag{8-7b}$$

α-Keto Alcohol Formation

$$CH_3\overset{\overset{\displaystyle O}{\|}}{C}COOH + CH_3\overset{\overset{\displaystyle O}{\|}}{C}—H \longrightarrow CH_3\underset{\underset{\displaystyle OH}{|}}{\overset{\overset{\displaystyle O}{\|}}{C}}—CHCH_3 + CO_2 \tag{8-7c}$$

All TPP reactions discovered thus far are analogous in that the carbon-carbon bond adjacent to a ketone (usually a keto-acid bond) is cleaved. This leads to the generation of a stable product whose nature depends on the particular enzyme involved.

The site of coenzyme function was quite convincingly demonstrated by Breslow[17,18] to be the C2 position of the thiazolium ring when he found that this position rapidly exchanges with D of D_2O under mild conditions. This implied that the anion at this C2 is very easily formed and that the mechanisms of TPP reactions are very much like typical nucleophilic carbonyl addition reactions where the nucleophile is the TPP carbanion at C2. This would lead to intermediate forms **XIIa** and **XIIb** as the first adduct. Depending on the enzyme, one could then observe oxidative or nonoxidative break-

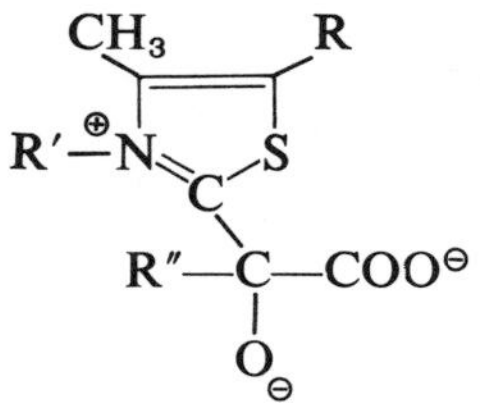

XIIa 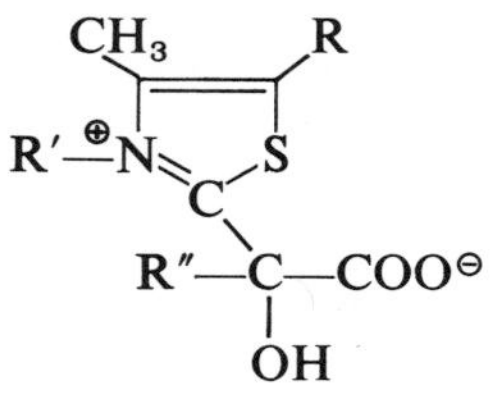   XIIb

down of **XII** or one could observe transfer to a suitable acceptor (α-keto alcohol

formation). Formally, then, TPP reactions involve the transfer of $RC\overset{\displaystyle O}{\diagup}$

to hydroxide, hydrogen, or RCHO.

b. Lipoic Acid

Lipoic acid is also involved in the transfer of R—C $\overset{\displaystyle O}{\underset{\diagdown}{\diagup}}$ groups. The process in which it participates is usually one of oxidative decarboxylation, which overall requires a complex multienzyme system. Lipoate is generally found to be a cofactor in the interaction between a TPP adduct and co-enzyme A (CoASH, see below). The lipoic acid is protein-bound through an amide linkage to its pentanoyl side chain carboxyl group. Thus Eq. 8-8 schematizes a typical lipoate reaction. It is a carrier coenzyme in that it is

$$(8-8)$$

not regenerated at the completion of its coenzymatic function but requires a separate step for this. Its mechanism appears to involve simple nucleophilic attack of the TPP-adduct anion on the bound lipoate to generate the S-acyl reduced lipoate, followed by transthiolation with coenzyme A.

c. Coenzyme A (CoASH)

Structurally, coenzyme A is one of the more complex of the organic coenzymes. It was recognized early that there is an essential organic cofactor in many tissue extracts for the acetylation of choline and sulfanilamide,[19] and this coenzyme of acetylation became commonly termed coenzyme A. Due to its structural complexity, the functional portion of the molecule was not recognized until Lynen isolated S-acetyl-CoA (**XIII**) from respiring yeast,

$$R—CH_2S—\overset{\displaystyle O}{\overset{\|}{C}}CH_3$$
XIII

demonstrating that the thiol group is the active acetyl carrier.[19] A great amount of effort has been devoted to the understanding of CoASH reactions. This coenzyme appears to be necessary in virtually all metabolic sequences involving the activation of, or reation at, a carboxylate group. The activation occurs through the generation of the S-acyl derivative, as outlined in Eq. 8-9. Thus CoASH is crucially required for acyl group transfer reactions in biosynthesis.

$$\underset{\text{O}}{\overset{\text{O}}{\parallel}}$$
$$R\overset{\text{O}}{\overset{\parallel}{C}}\text{---Donor} + \text{CoA---SH} \longrightarrow \text{Donor} + R\overset{\text{O}}{\overset{\parallel}{C}}\text{---SCoA} \xrightarrow{\text{Acceptor}}$$

$$R\overset{\text{O}}{\overset{\parallel}{C}}\text{---Acceptor} + \text{CoASH} \quad (8\text{-}9)$$

XIV

Once formed, a thiol ester such as **XIV** can react in one of three ways. First, nucleophilic or base-catalyzed attack at the carbonyl carbon may occur, generating the acyl-nucleophile + CoA (Eq. 8-10a). Second, the

$$\textbf{XIV} + \text{BH} \longrightarrow R\overset{\text{O}}{\overset{\parallel}{C}}\text{---B} + \text{CoASH} \qquad (8\text{-}10a)$$

methylene protons α to the carbonyl carbon of the thiol ester are somewhat acidic; hence that carbon can act as a carbanion under the proper circumstances, leading to nucleophilic displacement reactions or carbonyl addition reactions (Eqs. 8-10b, 8-10b′). Third, a structurally appropriate derivative

$$CH_3\overset{\text{O}}{\overset{\parallel}{C}}\text{---S---CoA} + \left\{ \begin{array}{c} CH_3\overset{\text{O}}{\overset{\parallel}{C}}\text{---SCoA} \\ \Updownarrow \\ \underset{OH}{CH_2\!=\!\overset{\mid}{C}\text{---SCoA}} \\ {}_{-H^{\oplus}}\Updownarrow{}_{+H^{\oplus}} \\ {}^{\ominus}CH_2\text{---}\overset{\text{O}}{\overset{\parallel}{C}}\text{---SCoA} \end{array} \right\} \longrightarrow CH_3\overset{\text{O}}{\overset{\parallel}{C}}CH_2\overset{\text{O}}{\overset{\parallel}{C}}\text{---SCoA} + \text{CoASH}$$

$$(8\text{-}10b)$$

$$R\overset{\text{O}}{\overset{\parallel}{C}}\text{---R}_1 + \quad \xrightarrow{+H^{\oplus}} \quad R\overset{OH}{\underset{R_1}{\overset{\mid}{\underset{\mid}{C}}}}\text{---CH}_2\text{---}\overset{\text{O}}{\overset{\parallel}{C}}\text{---SCoA}$$

$$(8\text{-}10b')$$

may undergo a reversible addition/elimination reaction about a carbon-carbon double bond (Eq. 8-10c). Enzymatic reactions corresponding to all three classes are known in biochemistry.[19] Jaenicke and Lynen[19] provided

$$R—\underset{\underset{Y}{\overset{\overset{X}{|}}{|}}}{CH}—CH—\overset{\overset{O}{\|}}{C}—SCoA \rightleftharpoons XY + RCH=CH—\overset{\overset{O}{\|}}{C}—SCoA \qquad (8\text{-}10c)$$

over 60 examples of CoASH reactions representative of all three classes mentioned here. Table 8-2 lists just one example from each class.

TABLE 8-2   Examples of Coenzyme A Reactions

Class	Enzyme	Reaction
8-10a	Choline acetylase	$CH_3COSCoA$ + Choline $\rightleftharpoons$ Acetyl-choline + CoASH
8-10b	β-Ketoacyl thiolase	Acetyl-CoA + Acetyl-CoA $\rightleftharpoons$ Acetoacetyl—CoA + CoASH
8-10c	Butyryl dehydrogenase	Butyryl—CoA + FAD $\rightleftharpoons$ FADH$_2$ + Crotonyl—CoA

d. Biotin

Biotin has long been recognized as an essential coenzyme in reactions of CO_2 transfer. Many CoASH reactions are ATP-dependent carboxylations involving a coupled system of enzymes—a carboxyl transferase, a biotin carboxylase, and a small "carboxyl carrier protein."[20] This carboxyl carrier protein is a small protein with biotin covalently attached through an amide linkage between its side chain carboxyl group and a protein amino group. It is carboxylated by biotin carboxylase; the carboxybiotin protein then reacts with the carboxyl transferase to transfer its CO_2 to the acceptor, which is usually an acyl-SCoA derivative. This is schematized in Eq. 8-11.

$$\text{(structure)}\ (CH_2)_4CONHPr + ATP + HCO_3^{\ominus} \xrightarrow{\text{biotin carboxylase}} ADP + Pi + \text{(structure, XV)}$$

$$(8\text{-}11)$$

$$XV + CH_3\overset{\overset{O}{\|}}{C}—SCoA \rightleftharpoons \text{(structure)} + {}^{\ominus}OOC—CH_2—\overset{\overset{O}{\|}}{C}—SCoA$$

In structure **XV** and in Table 8-1, the site of carboxylation of biotin has been left indeterminate. Although N_1-carboxybiotin (**XVIa**) has been isolated from biological systems,[4,20] Bruice and co-workers have provided compelling arguments that the species functional in carboxylation reactions is actually the O-carboxyderivative **XVIb**.[21-23] If **XVIb** is the active carboxylating

        XVIa   **XVIb**

species, attempts to isolate it would be expected to cause rearrangement to the thermodynamically more stable **XVIa**. Thus the isolation of **XVIa** does not prove that it is the active carboxylating species. Indeed, the expected lability of **XVIb** suggests that if it could be stabilized in some way by the carboxyl carrier protein until it contacts the carboxyl transferase, then **XVIb** could be quite effective at transferring its CO_2.

e. Tetrahydrofolic Acid (FH$_4$)

Tetrahydrofolic acid is involved in a variety of reactions involving the transfer of one carbon atom fragment at a variety of levels of oxidation, as indicated in Table 8-1. The reader is directed to reference 4 for a more complete discussion of this coenzyme and other leading references.

f. Cobalamins (Vitamin B$_{12}$)

Structurally, this coenzyme is perhaps the most complex coenzyme in nature. Its structure is shown in Figure 8-3. Thus far, only six reactions have been found to occur in mammalian systems that require B_{12} coenzyme; these are listed in Table 8-3.[24] Abeles has pointed out that these reactions have a common feature—in all of the reactions of Table 8-3, an X group and a hydrogen atom exchange places and there is no exchange of H with the medium. This is shown in Eq. 8-12. In the glutamate mutase case in Table 8-3, X is $COO^\ominus$. In example 2, $X = -\overset{\overset{\displaystyle O}{\|}}{C}-SCoA$. In examples 3 and 4,

$$-\underset{\underset{\displaystyle H}{|}}{\overset{\overset{\displaystyle |}{}}{C_1}}-\underset{\underset{\displaystyle X}{|}}{\overset{\overset{\displaystyle |}{}}{C_2}}-\ \xrightarrow{\ B_{12}\text{-enzyme}\ }\ -\underset{\underset{\displaystyle X}{|}}{\overset{\overset{\displaystyle |}{}}{C_1}}-\underset{\underset{\displaystyle H}{|}}{\overset{\overset{\displaystyle |}{}}{C_2}}- \qquad (8\text{-}12)$$

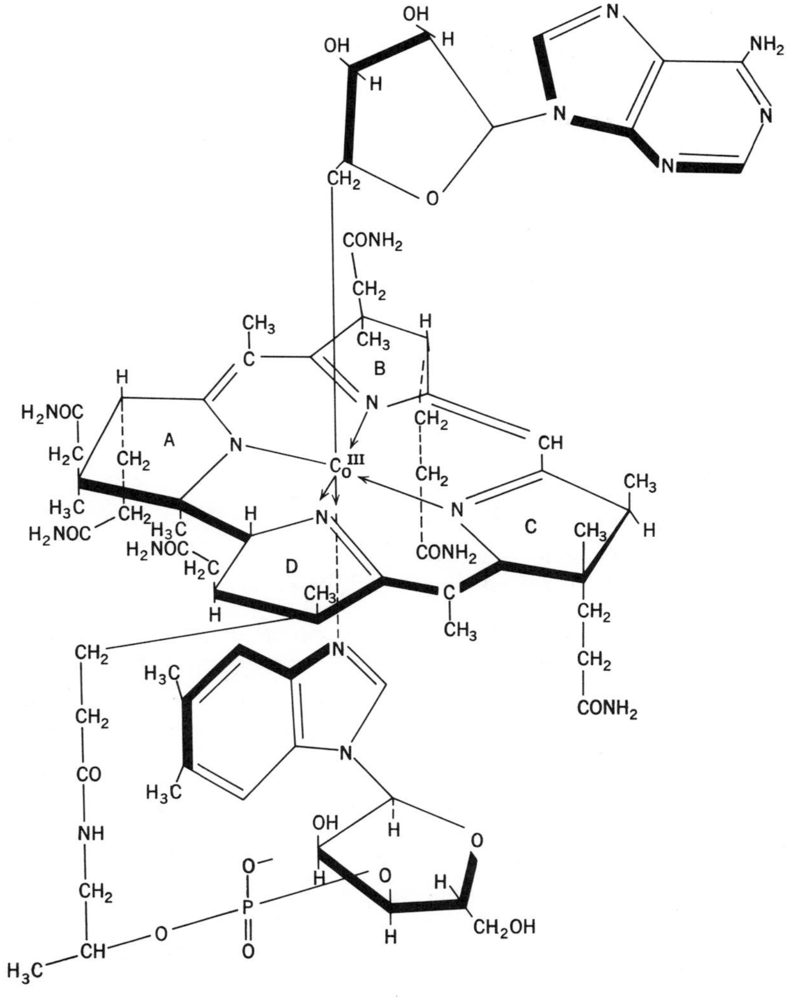

FIGURE 8-3. Structure of vitamin B_{12} coenzyme.

$X = OH$, but in these cases the first product will be the hydrated form of the aldehyde, which will rapidly dehydrate, liberating the product aldehyde.

During the enzymic reactions, the C5-deoxyadenosine-cobalt bond is ruptured and replaced by a bond between the cobalt and the substrate. The substrate transfers a hydride ion to the C5 carbon of the C5 deoxyribose during this process. This is shown in Eq. 8-13 for the diol dehydrase reaction. The formation of the cobalt-substrate bond is followed by rearrangement to

TABLE 8-3 Reactions Requiring Vitamin B_{12} Coenzymes

---

1. Glutamate mutase

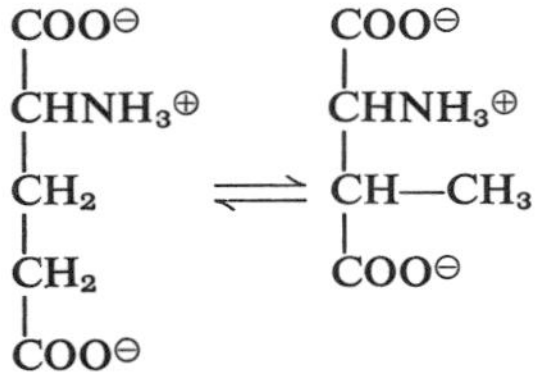

$$
\begin{array}{cc}
COO^{\ominus} & COO^{\ominus} \\
| & | \\
CHNH_3^{\oplus} & CHNH_3^{\oplus} \\
| & | \\
CH_2 & \rightleftharpoons \quad CH—CH_3 \\
| & | \\
CH_2 & COO^{\ominus} \\
| & \\
COO^{\ominus} &
\end{array}
$$

2. Methylmalonyl-CoA isomerase

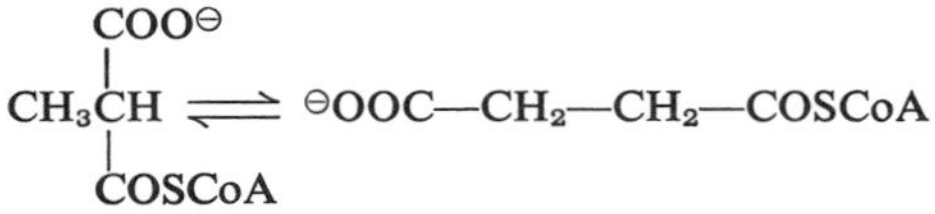

$$
\begin{array}{c}
COO^{\ominus} \\
| \\
CH_3CH \quad \rightleftharpoons \quad {}^{\ominus}OOC—CH_2—CH_2—COSCoA \\
| \\
COSCoA
\end{array}
$$

3. Glycerol-dehydrase

$$HOCH_2—CHOH—CH_2OH \longrightarrow CH_2OH—CH_2—CHO$$

4. Dioldehydrase

$$CH_3—CHOH—CH_2OH \longrightarrow CH_3—CH_2—CHO$$
$$CH_2OH—CH_2OH \longrightarrow CH_3—CHO$$

5. Ethanolamine deaminase

$$H_3N^{\oplus}CH_2—CH_2OH \longrightarrow CH_3—CHO + NH_4^{\oplus}$$

6. β-Lysine mutase

$$CH_2NH_3^{\oplus}(CH_2)_3CHNH_3^{\oplus}COO^{\ominus} \longrightarrow CH_3—CHNH_3^{\oplus}—(CH_2)_2—CHNH_3^{\oplus}COO^{\ominus}$$

---

intermediate **XVII**, which then breaks down to product and regenerated coenzyme.[24] Although this has not been rigorously proven, it fits the known kinetic and isotope exchange data.

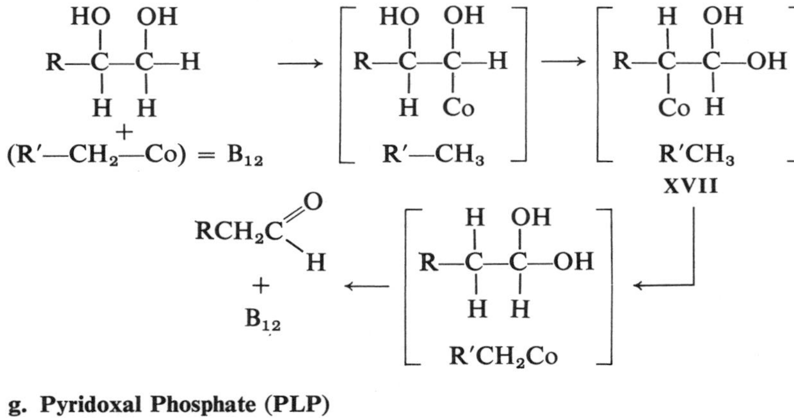

$$(8\text{-}13)$$

g. Pyridoxal Phosphate (PLP)

The nutritional factor vitamin B_6 was known in 1934 to be involved in protein metabolism. Chemical studies showed that its structure was 3-hydroxy-

4,5-di(hydroxymethyl)-2-methylpyridine (**XVIIIa**), called pyridoxol (PO).

CH$_2$OH

HOH$_2$C ⟨⟩ OH

N CH$_3$

XVIIIa
PO

CHO

HOH$_2$C ⟨⟩ OH

N CH$_3$

XVIIIb
PL

CHO

H$_2$O$_3$POH$_2$C ⟨⟩ OH

N CH$_3$

XVIIIc
PLP

CH$_2$NH$_2$

HOH$_2$C ⟨⟩ OH

N CH$_3$

XVIIId
PA

It was later found that other derivatives from the metabolism of PO were more effective, with PLP being recognized as the primary active form of the coenzyme.[4]

The great majority of PLP reactions involve metabolic transformations of amino acids. The types of amino acid reactions that require a PLP cofactor are listed in Table 8-4.

In considering how PLP functions in these various reactions, it is appropriate to consider how the coenzyme might be bound to its protein. Due to the multiplicity of functional groups on the pyridine rings of structures **XVIIIa** to **XVIIId**, a variety of possible mechanisms exists. Thus the phosphate group (PLP), the protonated amino group (PA), and a possibly ionized aromatic OH group might be bound by salt linkages. Although the last possibility seems unlikely, the first two could be quite important. The hydroxy groups (PO, PL, PA) might be held through hydrogen bonds to protein groups, and the aldehyde (PL, PLP) could be held either by hydrogen bonding or by covalent attachment via Schiff base formation. At this point definitive studies on the role of the methyl and aromatic hydroxy group have not been done. The phosphate role is also not well understood but is quite likely involved in a salt linkage with the enzyme protein.[25] Its possible contribution to the catalytic mechanism will be discussed later.

The aldehyde group has been shown in a number of enzymes to be crucial to PLP function through the formation of a Schiff base with the ε-amino group of a lysine residue on the protein (see Chapter 7). To name just three examples, borohydride reduction of the protein-PLP complex has allowed the eventual isolation of pyridoxyllysine from aspartate aminotransferase, bacterial glutamic decarboxylase, and threonine dehydrase.[25]

TABLE 8-4 Representative Amino Acid Reactions of Pyridoxal Phosphate Systems

Type	Reaction

Racemization

$$CH_3-\underset{\underset{NH_2}{|}}{\overset{\overset{H}{|}}{C}}-COOH \rightleftharpoons H_2N-\underset{\underset{CH_3}{|}}{\overset{\overset{H}{|}}{C}}-COOH$$

Transamination

$$HOOCCH_2CH_2\underset{\underset{NH_2}{|}}{CH}COOH + HOOC\overset{\overset{O}{\|}}{C}CH_2COOH \rightleftharpoons$$

$$+ HOOCCH_2CH_2\overset{\overset{O}{\|}}{C}COOH + HOOC-\underset{\underset{NH_2}{|}}{CH}CH_2COOH$$

Decarboxylation

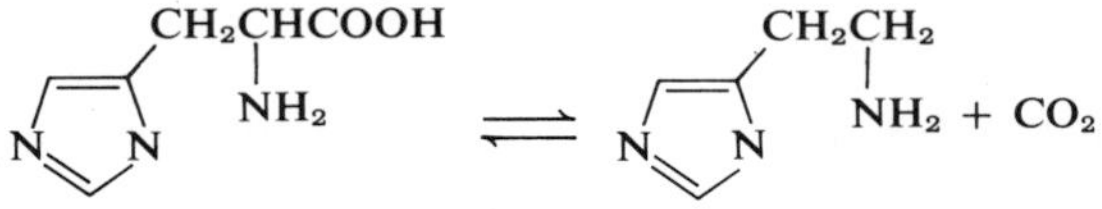

$$\underset{N\diagdown N}{\diagup}CH_2\underset{\underset{NH_2}{|}}{CH}COOH \rightleftharpoons \underset{N\diagdown N}{\diagup}CH_2\underset{\underset{NH_2}{|}}{CH_2} + CO_2$$

α-β-Elimination

$$HOCH_2\underset{\underset{NH_2}{|}}{CH}COOH \rightleftharpoons CH_2{=}\underset{\underset{NH_2}{|}}{CH}-COOH \longrightarrow CH_3\overset{\overset{}{}}{C}COOH \atop O$$

Synthesis of amino acids

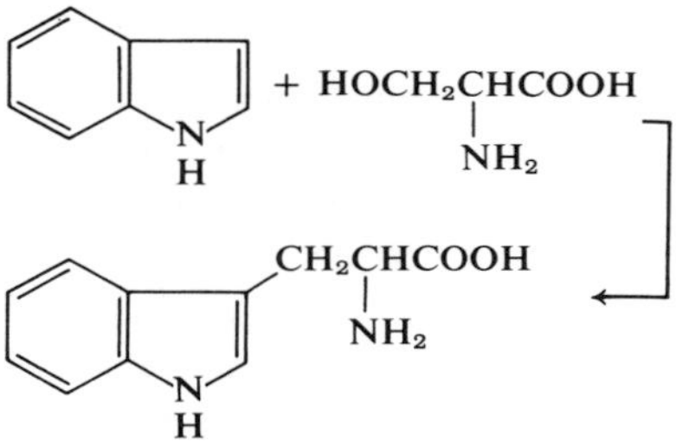

Model system studies for PLP function have proven extremely useful in elucidating the mechanisms of reactions requiring PLP. These studies have utilized higher temperatures and polyvalent metal ions along with PL or an analogous aldehyde, and even though the rates of the model reactions are sometimes as much as 10^6 less rapid, the transformations occurring with substrates of PLP requiring enzymes are essentially the same as the enzymic transformations. This implies that essentially the same pathways for product generation may exist for both the enzyme and the model.

From model system studies and from the knowledge of Schiff base formation between PLP and proteins, one is led to ask the purpose of Schiff base formation. In 1962, Cordes and Jencks[26] showed that the reaction of an amine with a Schiff base (transimination reaction) was some 30 times faster than Schiff base formation. Thus if an amino acid substrate is to form a Schiff base intermediate with the aldehyde of PLP, it can do so more readily through a transimination reaction than by reacting directly with the aldehyde. (This is reminiscent of the energy transfer function of ATP; see Eq. 8-6.)

The PLP-like reactions of model systems do not occur unless the polyvalent metal ion is present. This has led to the following description of the models, and, by analogy, for the PLP requiring enzymes (Scheme 8-1). In the models, M^{+3} substitutes only poorly for the enzyme, since it cannot form a Schiff base for transimination with a substrate and since it cannot bring about any manifestation of specificity.

The central structure in these studies is **XIX**. As shown, it has three susceptible bonds and the path of an enzymic reaction will ultimately depend on which bond, *a*, *b*, or *c*, breaks. If bond *a* breaks, the reaction may proceed in one of three ways—all through structure **XX**. Hydrolysis of **XX** leads to the transamination product, shown here as α-keto-β-hydroxypropionic acid; α-β-elimination leads to pyruvate from **XXI** since the initially formed aminoacrylate is hydrolytically unstable. The reversal of **XX** would lead to racemization as long as the enzyme cannot distinguish one side of the central carbon atom from the other. It is certainly true for the model system that reversal of formation of **XX** will lead to racemization of the initial amino acid substrate. The delicate stereospecificity of hydrogen addition in the $NAD^\oplus$ systems discussed earlier makes it seem possible that an enzymic hydrogen addition to **XX** might also be stereospecific, which could lead not to racemization but to either inversion or retention of configuration of the amino acid substrate. Since enzymic racemization reactions (racemases) have been observed,[27] either some pyridoxal enzymes behave differently from $NAD^\oplus$ enzymes with regard to the stereochemistry of hydrogen addition, or the racemization reaction is not really the prime function of the so-called "racemase."

If bond *b* breaks, one observes decarboxylation, and if *c* breaks, one observes deformylation (in the serine case) as shown in structures **XXII** and **XXIII**, respectively.

Structure **XIX**, the aldimine form of the intermediate produced by reaction of PLP with an amino acid, is analogous to an aldimine formed between enzymic PLP and an amino acid substrate. Before its formation, though, the PLP exists as an aldimine formed from reaction with a protein amino group. It has already been mentioned that this form of the enzyme-cofactor complex should minimally be 30 times more reactive than the aldehyde.[26] This is due to the inherently greater susceptibility of the aldimine carbon atom to attack by a nucleophilic amine. Consistent with this is a mechanistic proposal of Hamilton[29] shown in Scheme 8-2. This scheme depicts the probable

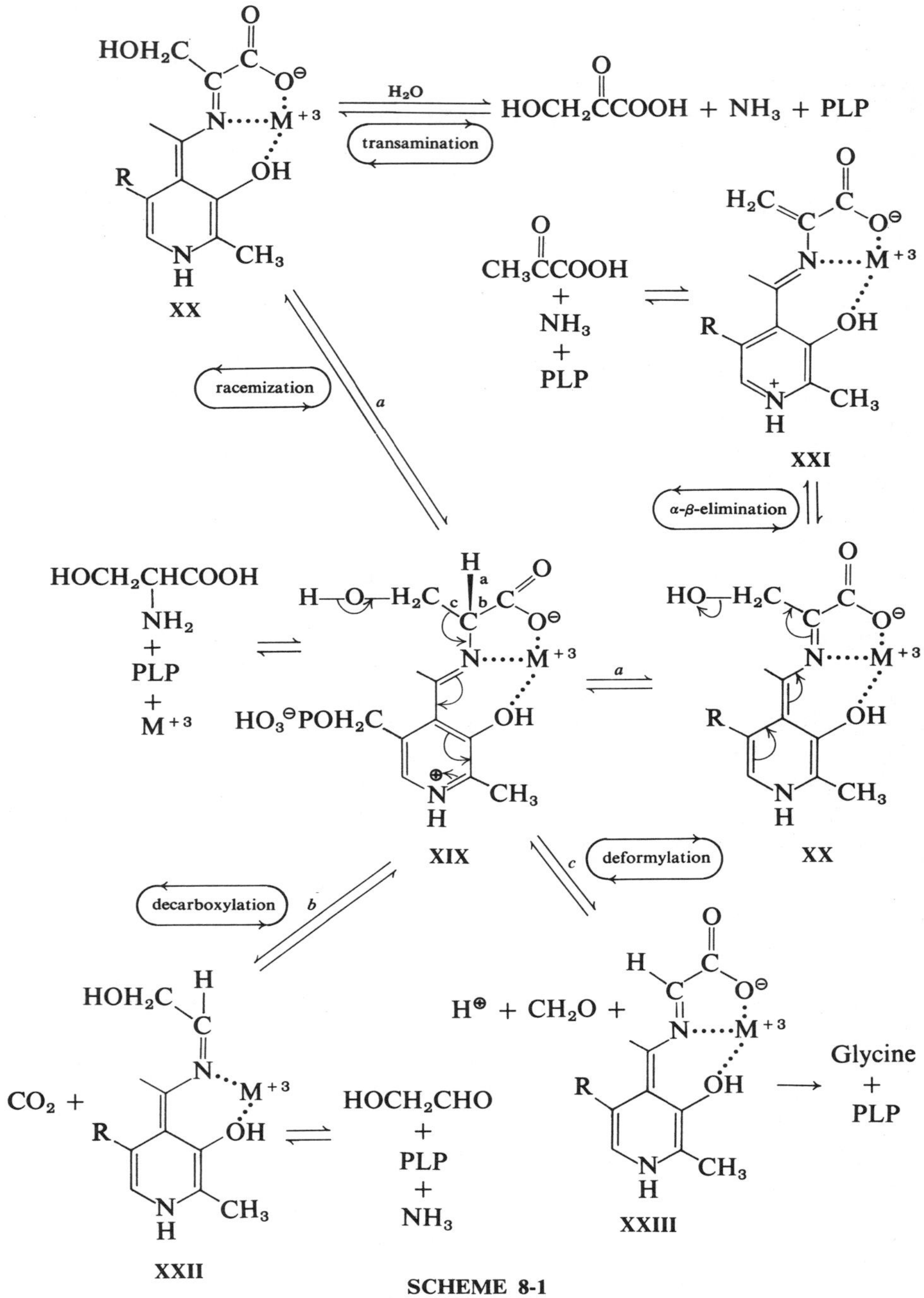

transamination
racemization
α-β-elimination
decarboxylation
deformylation
XX
XXI
XIX
XXII
XXIII
HOCH₂CCOOH + NH₃ + PLP
CH₃CCOOH + NH₃ + PLP
HOCH₂CHCOOH NH₂ + PLP + M+3
HOCH₂CHO + PLP + NH₃
H⊕ + CH₂O +
Glycine + PLP

SCHEME 8-1

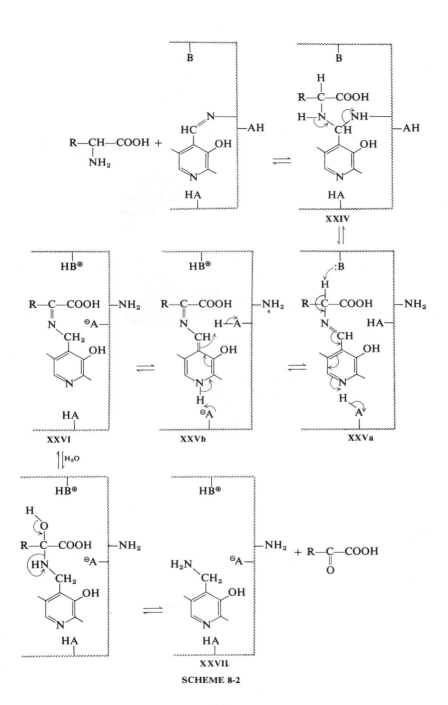

SCHEME 8-2

161

sequence of events in transamination reactions catalyzed by PLP-requiring enzymes. Here, instead of a metal ion, the PLP is attached to the enzyme as an aldimine, as previously mentioned. In the first step of the reaction the substrates' amino group, acting as a nucleophile, attacks the aldimine carbon to give **XXIV**. This then breaks down as shown to **XXV**, the transimination product. Then, with the participation of the appropriate acidic and basic groups on the enzyme, **XXVI** is formed. This hydrolyzes as shown to the pyridoxamine phosphate derivative of the enzyme, **XXVII**, and the product keto-acid. Regeneration of the PLP-enzyme can occur by reversal of this process with the appropriate keto-acid. An important point to mention about this reaction is that it is an oxidation reaction in which all hydrogen transfers are proton transfers and electrons shift through an extended network of conjugation. This will be considered further in Chapter 10.

Some PLP-requiring enzymes react with their substrates through a non-protonated aldimine. On the other hand, spectral studies have implied that some must be reacting through a protonated aldimine.[25] In these cases, and perhaps also in the nonprotonated aldimines to some extent, the phosphate group may be playing an important role in enhancing the reactivity of the aldimine carbon atom by increasing its electron deficiency. This role has been suggested for the C_1-phosphate group of aldolase substrates (see Chapter 11).

A particularly elegant example of the breakdown of an enzymic-PLP reaction into individual steps has been provided by Hammes and Fasella.[28] These authors used relaxation methods (temperature jump) to study the PLP-dependent enzyme, glutamic-aspartic transaminase. For this case, at least, the rate-determining step was shown to be the aldimine $\rightleftharpoons$ ketimine tautomerization, that is, structure **XXVa** going to structure **XXVb**.

Conclusion

Although many coenzyme systems have been studied at great length, few have been studied in as fine detail as the $NAD^\oplus$, CoASH, ATP, and PLP systems. Even for these, not all the pertinent questions have been answered. Clearly a more detailed understanding of ascorbic acid mediated enzymic reactions is needed; the same holds true for the quinones, cobalamins, and others. It also seems worthy of study to determine any possible coenzymatic function of dietary factors such as vitamin E, for which none is presently known. Such studies are both challenging and important.

References

1. M. Dixon and E. C. Webb, *Enzymes*, 2nd ed., Academic, New York, 1964, Chap. IX.
2. C. G. King, *Nutr. Rev.*, **26**, 33 (1968).

3. A. L. Tappel, *Am. J. Clin. Nutr.*, **23**, 1137 (1970).
4. H. R. Mahler and E. H. Cordes, *Biological Chemistry*, Harper and Row, New York, 1966, Chap. 4.
5. M. E. Pullman, A. San Pietro, and S. P. Colowick, *J. Biol. Chem.*, **206**, 129 (1954).
6. S. Chaykin, *Ann. Rev. Biochem.*, **36**, 149 (1967).
7. B. Vennesland and F. H. Westheimer, in *The Mechanism of Enzyme Action*, W. D. McElroy and B. Glass (Eds.), Johns Hopkins Press, Baltimore, 1954.
8. N. O. Kaplan and R. H. Sarma, *Pyridine Nucleotide Dependent Dehydrogenases*, Proc. Adv. Study Inst., H. Sund (Ed.), Springer Verlag, Berlin, 1970, p. 39.
9. S. Udenfriend, *Science*, **152**, 1335 (1966).
10. N. Stone and A. Meister, *Nature*, **194**, 555 (1962).
11. *Nutr. Rev.*, **25**, 183 (1967).
12. G. C. Willis, *Can. M.A.J.*, **77**, 106 (1957).
13. L. Pauling, *Proc. Nat. Acad. Sci., U.S.*, **67**, 1643 (1970).
14. L. Pauling, *Vitamin C and the Common Cold*, Freeman, San Francisco, 1970.
15. W. E. Blumberg, J. Peisach, B. A. Wittenberg, and J. B. Wittenberg, *J. Biol. Chem.*, **243**, 1854 (1968).
16. A. H. Neims and L. Hellerman, *Ann. Rev. Biochem.*, **39**, 867 (1970).
17. R. Breslow, *J. Am. Chem. Soc.*, **79**, 1762 (1957).
18. R. Breslow, *J. Am. Chem. Soc.*, **80**, 3719 (1958).
19. L. Jaenicke and F. Lynen, in *The Enzymes*, Vol. 3, P. D. Boyer, H. A. Lardy, and K. Myrback, (Eds.), Academic, New York, 1960, p. 3.
20. J. Knappe, *Ann. Rev. Biochem.*, **39**, 757 (1970).
21. T. C. Bruice and A. F. Hegarty, *Proc. Nat. Acad. Sci., U.S.*, **67**, 805 (1970).
22. A. F. Hegarty and T. C. Bruice, *J. Am. Chem. Soc.*, **92**, 6561, 6575 (1970).
23. R. F. Pratt and T. C. Bruice, *Biochemistry*, **10**, 3178 (1971).
24. R. H. Abeles, in *Bioinorganic Chemistry, Advances in Chemistry Series 100*, Am. Chem. Soc., Washington, D.C., 1971, p. 346.
25. P. Fasella, *Ann. Rev. Biochem.*, **36**, 185 (1967).
26. E. H. Cordes and W. P. Jencks, *Biochemistry*, **1**, 773 (1962).
27. A. E. Braunstein, in *The Enzymes*, Vol. 2, P. D. Boyer, H. A. Lardy, and K. Myrback (Eds.), Academic, New York, 1960, p. 113.
28. G. G. Hammes and P. Fasella, *J. Am. Chem. Soc.*, **84**, 4644 (1962).
29. G. A. Hamilton, in *Progress in Biorganic Chemistry*, Vol. 1, E. T. Kaiser and F. J. Kézdy (Eds.), Wiley, New York, 1971, p. 83.

PART TWO

Enzyme Mechanism Studies

The number of different enzyme-catalyzed reactions that have been identified to date is estimated at something over 1000. Among the enzymes responsible for the catalysis in each of these reactions, the number that has actually been isolated and purified to a degree suitable for studies directed toward mechanism elucidation is smaller by almost an order of magnitude. And among the enzymes upon which mechanism studies have been carried out to one degree or another, only a relative handful, perhaps 15 or 20, have been examined in enough detail to provide a firm basis for speculation concerning a complete mechanism of enzyme action. Even among these few for which reasonably complete mechanisms of action can be proposed, there is as yet not a single one whose proposed mechanism can be regarded as "established" or for which we are able to account in quantitative terms for the magnitude of catalysis. Yet the very fact that we are able experimentally to obtain data upon which good, firm, scientifically sound mechanistic hypotheses can be confidently advanced is exciting and important.

In the following chapters, we consider in some detail a few of the enzymes and enzyme-catalyzed reactions that have received intensive study and for which mechanisms can in fact be proposed. The enzymes and reactions considered in these chapters are not necessarily the best understood or the most important ones. Nor is it the purpose of these chapters to serve as reviews of the literature. Rather this portion of the book is to be regarded as a collection of case studies in which principles presented earlier may be seen applied to real enzymes in a systematic way.

CHAPTER 9

Hydrolytic Enzymes

Among all the many classes and subclasses of enzymes (see Appendix), some that catalyze hydrolysis reactions have received a seemingly disproportionate amount of study. Most of these intensively studied hydrolases are extracellular proteins that are readily isolated and purified. As such they were among the earliest enzymes available in great enough quantity and in high enough purity for systematic study. Furthermore, for the most part they are relatively simple, stable, low molecular weight enzymes, and the hydrolysis reactions they catalyze are not difficult to monitor experimentally.

Most of these well-known hydrolytic enzymes are proteases or protein-cleaving enzymes. These include the serine proteases such as chymotrypsin, trypsin, subtilisin, elastase and thrombin; the cysteine proteases such as papain, ficin, and bromelain; the low pH proteases such as pepsin and rennin; the metalloexopeptidases such as the carboxypeptidases and aminopeptidases; and the neutral metalloendopeptidases such as thermolysin and related enzymes of bacterial origin. All of these proteases catalyze the hydrolysis of peptide linkages. In nature we find them in the digestive tracts of animals where they catalyze the hydrolysis of food protein into dialyzable fragments, in plants where they catalyze the turnover of protein, and also in bacteria where they presumably function as digestive enzymes on proteinaceous substrates of nutrient media.

In this chapter we will probe fairly extensively into the enormous literature devoted to chymotrypsin, which is clearly the world's most studied enzyme. As such it serves as a prototype among enzymes in general, and numerous techniques that may be applied broadly in enzyme mechanism studies were developed originally in conjunction with chymotrypsin studies or have been evaluated using this enzyme as a test system. We will also have a look at carboxypeptidase A, a protease of particular interest since it is a metalloenzyme.

Although generally the best known hydrolases are found among the proteases, there are some exceptions, perhaps the most notable being lysozyme. Lysozyme, an enzyme found in large quantities in egg white, catalyzes the hydrolysis of glycosidic linkages in certain polysaccharide substrates. It was the first *enzyme* protein whose structure was determined by X-ray crystallography, and is an excellent case study illustrating the impressive power of that technique in enzyme investigation.

Chymotrypsin (EC 3.4.4.5)

a. Occurrence, Isolation, and Purification

Chymotrypsin is one of several proteolytic enzymes or *proteases* which function collectively in the mammalian small intestine. These proteases catalyze the hydrolytic cleavage of polypeptides from food protein into fragments small enough to dialyze out through the intestinal wall membranes. It is impressive to note that under the conditions of the small intestine, protein and polypeptide digestion in the presence of proteases proceeds some 10^5 times faster than it would in their absence. Thus digestion of a steak dinner without enzymes would take approximately 50 years instead of a few hours.

Chymotrypsin originates in the mammalian pancreas where it is produced in the form of an enzymatically inactive precursor, or *zymogen*, called chymotrypsinogen. The pancreas produces several such proteolytic zymogens, among which, in addition to chymotrypsinogen, are trypsinogen, procarboxypeptidase, and proelastase. Note that if these proteases were biosynthesized in active form, they might prematurely hydrolyze each other as well as the various enzymes of the cells in which they are produced.

Inactive chymotrypsinogen enters the intestine through the pancreatic duct. Once in the intestine, it encounters active trypsin and chymotrypsin, which activate the newly arrived zymogen by the specific hydrolysis of certain peptide bonds. This process of activation will be discussed in more detail later, but first let us briefly consider chymotrypsinogen itself.

Bovine chymotrypsinogen A (CTogen-A), an alkaline (pI = 9.1) globular protein of molecular weight 25,000, is readily obtained from an acidic extract of fresh beef pancreas by a procedure originally described by Kunitz[1] and outlined more recently by Wilcox.[2] After some initial fractionations of the extract with ammonium sulfate, CTogen-A can be easily crystallized and repeatedly recrystallized nearly to homogeneity from ammonium sulfate solutions or from aqueous ethanol. Fairly pure, crystalline CTogen-A is available commercially. If desired, it may be further purified by chromatography on CM-cellulose.

CTogen-A is converted to active chymotrypsin *in vitro* upon treatment of the zymogen with bovine trypsin at pH 7.6. The trypsin treatment results in the specific hydrolytic cleavage of the peptide bond between Arg-15 and Ile-16 in the CTogen amino acid sequence (see Table 9-1). The cleavage of this one bond gives rise to a form of the enzyme designated CTA_π, which is fully

TABLE 9-1   Amino Acid Sequence of Bovine Chymotrypsinogen A[a]

Cys	Gly	Val	Pro	5 Ala	Ile	Gln	Pro	Val	10 Leu	Ser	Gly	Leu	Ser	15 Arg	Ile	Val	Asn	Gly	20 Glu
Glu	Ala	Val	Pro	25 Gly	Ser	Trp	Pro	Trp	30 Gln	Val	Ser	Leu	Gln	35 Asp	Lys	Thr	Gly	Phe	40 His
Phe	Cys	Gly	Gly	45 Ser	Leu	Ile	Asn	Glu	50 Asn	Trp	Val	Val	Thr	55 Ala	Ala	His	Cys	Gly	60 Val
Thr	Thr	Ser	Asp	65 Val	Val	Val	Ala	Gly	70 Glu	Phe	Asp	Gln	Gly	75 Ser	Ser	Ser	Glu	Lys	80 Ile
Gln	Lys	Leu	Lys	85 Ile	Ala	Lys	Val	Phe	90 Lys	Asn	Ser	Lys	Tyr	95 Asn	Ser	Leu	Thr	Ile	100 Asn
Asn	Asp	Ile	Thr	105 Leu	Leu	Lys	Leu	Ser	110 Thr	Ala	Ala	Ser	Phe	115 Ser	Gln	Thr	Val	Ser	120 Ala
Val	Cys	Leu	Pro	125 Ser	Ala	Ser	Asp	Asp	130 Phe	Ala	Ala	Gly	Thr	135 Thr	Cys	Val	Thr	Thr	140 Gly
Trp	Gly	Leu	Thr	145 Arg	Tyr	Thr	Asn	Ala	150 Asn	Thr	Pro	Asp	Arg	155 Leu	Gln	Gln	Ala	Ser	160 Leu
Pro	Leu	Leu	Ser	165 Asn	Thr	Asn	Cys	Lys	170 Lys	Tyr	Trp	Gly	Thr	175 Lys	Ile	Lys	Asp	Ala	180 Met
Ile	Cys	Ala	Gly	185 Ala	Ser	Gly	Val	Ser	190 Ser	Cys	Met	Gly	Asp	195 Ser	Gly	Gly	Pro	Leu	200 Val
Cys	Lys	Lys	Asn	205 Gly	Ala	Trp	Thr	Leu	210 Val	Gly	Ile	Val	Ser	215 Trp	Gly	Ser	Ser	Thr	220 Cys
Ser	Thr	Ser	Thr	225 Pro	Gly	Val	Tyr	Ala	230 Arg	Val	Thr	Ala	Leu	235 Val	Asn	Trp	Val	Gln	240 Gln
Thr	Leu	Ala	Ala	245 Asn															

[a] Chymotrypsin consists of a single, unbroken chain of amino acid residues as indicated in the table beginning with a ½-cystine residue at the N-terminus and ending with a C-terminal asparagine. The protein has five disulfide bridges as follows: I–IV (residues 1–222), II–III (residues 42–58), V–IX (residues 136–201), VI–VII (residues 168–182), and VIII–X (residues 191–220).

active but highly susceptible to autolysis (self-digestion). CTA_π is normally transformed into CTA_δ by an autocatalytic action of active chymotrypsin, which removes the dipeptide Ser_{14}-Arg_{15}. Still further chymotryptic autolysis results in the loss of a second dipeptide, Thr_{247}-Asn_{248}, giving CTA_γ. Finally, CTA_γ may be converted to CTA_α by precipitating the activation mixture at pH 3 with ammonium sulfate and crystallizing at pH 4. The α and γ forms are identical in primary structure but have some slight differences in conformation and physical properties. Although all four of the forms of CTA mentioned above are active enzymes and are very similar in their specificities and modes of action, that α-form is the one that has been studied by far the most extensively because it is the most readily obtained free of contamination by the other forms. In all of our subsequent discussion concerning chymotrypsin, it will be implicit that we are dealing with bovine CTA_α unless specifically stated to the contrary.

Crystalline CTA_α of varying degrees of purity is available in quantity from commercial sources. Although crystalline, it always contains a significant amount of "inert protein," typically 10–15% by weight, which cannot be removed by repeated crystallizations or by any other means yet attempted. Low molecular weight impurities and small autolysis products also contaminate CTA_α preparations, and if they are not removed they may give rise to anomalous behavior in CTA_α studies. Gel filtration of a CTA_α preparation through a column of Sephadex G-25 removes these low molecular weight contaminants, giving a chymotrypsin solution that may be used with confidence for a day or two before slow autolysis again produces significant contamination.

b. General Catalytic Properties

We have already mentioned that the biological function of chymotrypsin is the catalysis of peptide bond hydrolysis for polypeptide substrates in the mammalian small intestine. It has a preferential specificity for peptide bonds in which the carboxylic carbonyl group belongs to one of the aromatic amino acids, L-phenylalanine, L-tyrosine, or L-tryptophan.[3] However, it can and does catalyze hydrolyses of other peptide bonds at a much slower rate.

As is the case in general with most enzymes, chymotrypsin activity varies as a function of pH. The optimal pH for CTA is between 7.5 and 9.0. Above and below this optimal range, CTA activity falls off sharply. We shall have more to say about the mechanistic implications of this pH dependence later.

In addition to its ability to catalyze polypeptide hydrolysis, chymotrypsin is a more or less effective catalyst for the hydrolysis of amide and ester linkages in a wide variety of synthetic substrates, some of which bear little structural resemblance to naturally occurring peptide substrates. Broad specificity of this sort is not unique to chymotrypsin. It is observed with a number of proteolytic enzymes and has been used to advantage in investigations designed to probe the structural characteristics of these enzymes' active sites (see Section e below). Furthermore, because simple synthetic substrates are generally much easier to obtain and work with than polypeptides, they have been used almost to the exclusion of polypeptides in kinetic and mechanism studies on chymotrypsin and other proteases on the assumption that in all important respects the enzyme interacts with them and with naturally occurring polypeptide substrates in the same manner.* Ester substrates have been especially useful in this regard, for as a rule their hydrolysis rates are more easily monitored than those of amides or peptides and they tend to react more rapidly than do amides of corresponding structure.

In the context of the preceding paragraph, it has become accepted to refer to chymotrypsin substrates as *specific substrates* if they are amides or esters of α-N-acylated aromatic amino acids. Those substrates which are not of this type are termed *nonspecific substrates. Natural substrates*—polypeptides in which cleavage occurs at the carboxyl function of an aromatic amino acid residue—are included within the broader specific substrate classification. The general area of substrate specificity will be discussed in greater detail for chymotrypsin below (Section e).

c. Assay

CTA preparations are most frequently assayed by kinetic methods. Several such methods have been described,[2] but one which is very commonly employed is a measurement of the rate of hydrolysis of N-acetyl-L-tyrosine ethyl ester (ATEE) at an initial concentration of $0.01M$ in $0.10M$ $CaCl_2$ at pH 8.6 and 25°C (Eq. 9-1). The reaction is followed by means of an automatic pH-stat titrator, which adds base to titrate the acid formed as a

$$CH_3\overset{O}{\overset{\|}{C}}NHCHC\overset{O}{\overset{\|}{-}}OC_2H_5 + H_2O \xrightarrow[\text{pH 8.6}]{\text{CTA}_\alpha} CH_3\overset{O}{\overset{\|}{C}}NHCHC\overset{O}{\overset{\|}{O}}{}^{\ominus} + HOC_2H_5 + H^{\oplus}$$

(9-1)

---

* This assumption appears to hold fairly well in the case of chymotrypsin, but it is not to be taken for granted that this is true for all proteases. For example, both carboxypeptidase A and papain have been found to have rather extensive active sites which "recognize" structural variations in polypeptide substrates several amino acid residues removed from the site of hydrolytic cleavage.

product in order to maintain the pH constant at 8.6. A stripchart recording is made of the volume of standard base added as a function of time, and this of course is a direct measure of the rate of the reaction in question.

The protein concentration employed is about 1 mg/ml. CTA_α concentration is generally estimated spectrophotometrically at 282 nm, at which wavelength $E_{1\,cm}^{1\%} = 20.5$. In terms of molar quantities (mol. wt. of $CTA_\alpha \simeq 25,000$), this concentration is about $4 \times 10^{-8}M$, which may be estimated using $\varepsilon_{282} = 5 \times 10^4 M^{-1}\,cm^{-1}$. It should be borne in mind that these concentration values obtained spectrophotometrically are only rough estimates of actual active CTA_α concentrations, since commonly used preparations contain up to 15% by weight of inactive protein.[2]

Under the conditions described, the rate of hydrolysis of ATEE is essentially constant for the first 25% of reaction, and an accurate velocity can be calculated from the slope of the linear trace on the recorder chart. Since

$$v_0 = \frac{k_{cat}[E]_0[S]_0}{K_{m(app)} + [S]_0} \qquad (9\text{-}2)$$

(see Chapter 4), the active enzyme concentration may be determined from the measured velocity of the assay and the known values of k_{cat} and $K_{m(app)}$,* which are 193 sec^{-1} and $7 \times 10^{-4}M$, respectively.[4] Thus, for $[S]_0 = 10^{-2}M$,

$$[E]_0 = \frac{v_0(K_{m(app)} + [S]_0)}{k_{cat}[S]_0} = v_0(5.55 \times 10^{-3}) \qquad (9\text{-}3)$$

Expressed in standard units of enzyme activity, the specific activity of CTA_α is 464 μmole of ATEE hydrolyzed per minute per milligram of enzyme under the conditions described above.

d. Reaction Pathway

It should be recalled that chymotrypsin is among the enzymes for which it has essentially been established that a covalent enzyme-substrate intermediate lies on the catalysis pathway (see Chapter 7). The simplest sequence of events consistent with the known behavior of this enzyme is indicated in Eq. 9-4, where ES is the usual noncovalent enzyme-substrate complex, or Michaelis complex, and ES' is the covalent acyl-enzyme intermediate. P_1 is the initially

$$E + S \underset{k_{-1}}{\overset{k_1}{\rightleftharpoons}} ES \xrightarrow{k_2} ES' \xrightarrow{k_3} E + P_2 \qquad (9\text{-}4)$$
$$\searrow$$
$$P_1$$

* These values for k_{cat} and $K_{m(app)}$ were determined using a CTA_α preparation whose active-site normality was accurately predetermined using a direct active-site titration procedure as described for chymotrypsin in Chapter 3. Such an active-site titration is superior to the rate assay method described here for the determination of CTA_α concentrations but is more difficult to carry out and requires substantially more enzyme.[5]

released amine or alcohol moiety of the amide or ester substrate and P_2 is the accompanying carboxylic acid moiety.

The steady-state kinetic expressions that apply to Eq. 9-4 were derived in Chapter 5, but for convenient reference we will restate them here:

$$v_0 = \frac{k_{cat}[E]_0[S]_0}{K_{m(app)} + [S]_0} \tag{9-5}$$

where

$$k_{cat} = \frac{k_2k_3}{k_2 + k_3} \tag{9-6}$$

and

$$K_{m(app)} = \frac{K_sk_3}{k_2 + k_3} \tag{9-7}$$

with

$$K_s = \frac{k_2 + k_{-1}}{k_1} \tag{9-8}$$

An impressive array of experimental evidence has been accumulated in support of the presumption that Eq. 9-4 represents at least a minimally adequate pathway for chymotrypsin-catalyzed hydrolyses.[6–8] An extensive discussion of this evidence cannot be undertaken here, but some of the more compelling aspects of it are well worth our attention, not only for their implications as regards the chymotrypsin problem, but also as examples of experimental approaches that may be applied in other enzyme studies when the involvement of covalent intermediates is suspected.

Early evidence for the applicability of Eq. 9-4 to chymotrypsin-catalyzed hydrolysis was provided by studies of Hartley and Kilby[9] with the non-specific chymotrypsin substrate, p-nitrophenyl acetate (p-NPA). The CTA_α-catalyzed hydrolysis of p-NPA at pH 8 was monitored spectrophotometrically by following the rate of appearance of bright yellow p-nitrophenolate ion. For just a few seconds after the reaction was initiated, nitrophenol was produced very rapidly up to the point where about 1 mole had appeared per mole of enzyme present.* Then the production of nitrophenol rather abruptly settled down to a slow constant rate until p-NPA was exhausted. This behavior is illustrated in Figure 9-1. Hartley and Kilby interpreted their results in terms of Eq. 9-4 with $P_1 = $ p-nitrophenol, $P_2 = $ acetate, and

* This is just the sort of behavior that may be exploited in so-called active-site titration procedures for the determination of absolute active enzyme concentrations.[5] (See Chapters 3 and 5.)

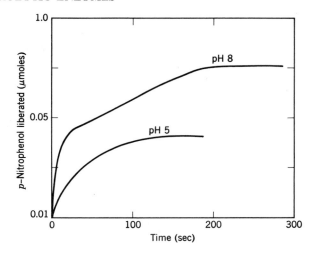

FIGURE 9-1. Time-course curves for the reaction of chymotrypsin with *p*-nitrophenyl-acetate. At pH 8 there is a rapid "burst" of *p*-nitrophenol followed by a zero-order turnover until the substrate supply is exhausted. At pH 5, the liberation of *p*-nitrophenol is initially much slower and stops completely when all of the enzyme is acetylated.

$k_2 \gg k_3$. The initial "burst" of nitrophenol coincided with the fast, presteady-state acetylation of a nucleophilic group at the enzyme-active site. The ensuing slow, steady-state production of P_1 resulted during "turnover" as the acyl-enzyme intermediate hydrolyzed and was then quickly reacetylated over and over again. Other investigators[10] later found that at lower pH, the turnover rate following the initial burst was so slow (see Figure 9-1) that the acetyl-enzyme produced could be isolated and studied. The trimethylacetyl derivative of chymotrypsin, prepared by the reaction of CTA_α with the *p*-nitrophenyl ester of trimethylacetic acid at pH 5, has even been crystallized.[11]

These isolated acetyl-chymotrypsin derivatives are found to be enzymatically inactive, but full activity (toward assay substrates such as ATEE) may be regenerated upon removal of the acetyl group by exposing the acetyl enzyme to higher pH or even at pH 5 by exposure to nucleophilic reagents more potent than water (such as hydroxylamine). It is important to note that both the acetylation and the deacetylation processes require that the enzyme remain in its native conformation. Thus denaturation of chymotrypsin with $8M$ urea totally inactivates the enzyme toward *p*-NPA (as well as toward specific substrates). Moreover, urea denaturation of an isolated acetyl chymotrypsin prevents deacetylation between pH 5 and 8 either by water or by hydroxylamine.[12]

A number of other analogous inactive acyl-enzyme intermediates have been prepared and studied by various research groups studying chymotrypsin.[6,7] Even before the p-NPA studies mentioned earlier, it was established that chymotrypsin can be totally inactivated by reaction with one mole of an alkyl phosphate such as diisopropylphosphorofluoridate (DFP).[13] The reaction is shown in Eq. 9-9. Unlike p-NPA, alkyl phosphates like DFP and certain alkyl sulfonyl fluorides such as phenylmethanesulfonyl fluoride[14] (see

$$E-H + (RO)_2-\overset{\overset{\displaystyle O}{\|}}{P}-F \longrightarrow E-\overset{\overset{\displaystyle O}{\|}}{P}-(OR_2) + HF \qquad (9-9)$$

Chapter 3) are not really substrates for chymotrypsin, since even under optimal conditions of pH, deacylation (actually dephosphorylation or desulfonylation) fails to occur. That is, the value of k_3 in Eq. 9-4 for these so-called quasi-substrates[15] is effectively zero except in the presence of potent nucleophilic acyl-group acceptors such as hydroxylamine, which can effect at least partial regeneration of active chymotrypsin.[16]

As with p-NPA, it is to be noted that these quasi-substrates do not react with denatured chymotrypsin, which suggests that the normal substrate-recognizing capabilities of intact chymotrypsin active sites are essential to the reaction and that it must indeed occur at the active site itself. It has in fact been found upon mild enzymatic degradation of both the diisopropyl-phosphoryl chymotrypsin[17] and the trifluoroacetyl chymotrypsin[18] derivatives that the phosphorylated or acetylated functional group is the hydroxyl OH of a unique serine residue of the enzyme polypeptide chain. Sequence studies[19] identify this unique residue as serine 195 (see Table 9-1).

It is appropriate to point out at this stage in our discussion that chymotrypsin is just one of a number of well-characterized enzymes (many of them proteases) which are inactivated by DFP as in Eq. 9-9. Others include[20,21] trypsin, thrombin, elastase, alkaline phosphatase, butyrylcholinesterase, acetylcholinesterase, liver aliesterase, subtilisin, mold protease, and phosphorylase. In each of these enzymes it has been established that the residue attacked by DFP is a unique reactive serine, presumably at the active site. Indeed it has for some time been generally accepted that inactivation by DFP (or other similar alkyl phosphates) is the definitive criterion for recognition of a "serine enzyme." Furthermore, it is widely suspected that in all of these serine enzymes, despite their diversity in substrate specificity, the serine hydroxyl group plays a vital role as a nucleophile in the mechanism of catalysis (see Chapter 7).

The isolation and characterization of acetyl chymotrypsins makes it virtually certain that the chymotrypsin-catalyzed hydrolysis of p-nitrophenyl acetates follows Eq. 9-4 and the inactivation of the enzyme by DFP and other such quasi-substrates is clearly a related phenomenon. Unfortunately, however, these compounds which give rise to isolatable intermediates are not *specific* substrates for chymotrypsin; compounds that are specific substrates do not give rise to isolatable intermediates. Nevertheless, there is excellent spectrophotometric and kinetic evidence that the acyl-enzyme pathway is valid in general for specific as well as for nonspecific chymotrypsin substrates.[6,7]

For example, the buildup and breakdown of an acyl-enzyme intermediate has been observed spectrophotometrically in the chymotrypsin-catalyzed hydrolysis of N-acetyl-L-tryptophan methyl ester at low pH. "Bursts" of *p*-nitrophenol such as that described earlier for *p*-NPA have been seen at low pH in the chymotrypsin-catalyzed hydrolysis of the *p*-nitrophenyl esters of N-acetyl-L-tryptophan and N-benzyloxycarbonyl-L-tyrosine, and they are identical within experimental error. This is consistent with the rate-determining deacylation of a common acyl-enzyme intermediate following a fast acylation step in which the alcohol leaving group is displaced by the enzyme (Eq. 9-4 with $k_2 \gg k_3$). Other evidence gathered from steady-state kinetic data with varying pH, varying substrate structure, and varying acyl group acceptors such as alcohols and amines in place of the usual water all provides a self-consistent picture for chymotrypsin catalysis, which agrees with the acyl-enzyme hypothesis.[6,7]

Some of the latest and most convincing evidence for the applicability of Eq. 9-4 to chymotrypsin-catalyzed hydrolyses is provided by spectrophotometric kinetic studies under presteady-state conditions using stopped-flow and temperature jump techniques and intensely chromophoric substrates and inhibitors. The kinetic equations that apply to such studies have been outlined in detail in Chapter 5. The use of the chromophoric furylacryloyl group **I** in chymotrypsin studies was introduced by Bernhard in 1965.[23]

I

Bernhard and Gutfreund[24] report that the CTA$_\alpha$-catalyzed hydrolysis of the nonspecific substrate β-2-furylacryloylimidazole **II** (Eq. 9-10) may be monitored spectrophotometrically near the wavelength at which the substrate **II** and the product **IV** have equivalent molar absorptivities (i.e., the substrate-product isosbestic point). At such a wavelength, changes in absorbance during the time course of the reaction can be attributed to the buildup and decomposition of transient species (such as **III**) which have different molar absorptivities. Using stopped-flow techniques these investigators were able

$$\text{II} + \text{Enz-OH} \underset{}{\overset{K_s}{\rightleftharpoons}} \text{ES} \xrightarrow{k_2}$$

$$\text{III} + \text{imidazole} \xrightarrow{k_3} \qquad (9\text{-}10)$$

$$\text{IV} + \text{Enz-OH}$$

to directly observe and measure both the (fast) acylation and (slow) deacylation rates of Eq. 9-10.

In a different set of experiments involving the dye proflavine, which is a competitive inhibitor for CTA_α and whose visible spectrum is perturbed when the dye-enzyme complex is formed, reaction 9-10 was followed by monitoring the rates at which the dye-enzyme complex disappeared and then reappeared. These rates were found to be identical to the rates of acyl-enzyme appearance and disappearance, respectively, in the other experiments. Finally it was ascertained that with the substrate β-2-furylacryloyl-p-nitrophenolate V, the rate of appearance of chromophoric p-nitrophenol corresponded identically with the rate of appearance of furylacryloylchymotrypsin

(III) as required by Eq. 9-4; furthermore, the overall (steady-state) rates of hydrolysis of the imidazolyl (II) and p-nitrophenyl (V) derivatives were identical, as predicted by Eq. 9-4 for the case where k_3 (deacylation) is rate limiting and the reactions being compared have a common acyl-enzyme intermediate (III).

These studies by Bernhard and his co-workers establish quite convincingly that Eq. 9-4 as written is consistent with the catalytic behavior of chymotrypsin toward II and V. However, these again are nonspecific substrates and are hydrolyzed much more slowly than optimal specific substrates. Clearly, similar investigations with specific substrates for chymotrypsin are called for.

Such studies have been carried out by Hess and his co-workers.[25] Using the proflavine displacement method mentioned earlier and the stopped-flow technique, Brandt, Himoe, and Hess[26] were able to determine the individual rate and equilibrium constants of Eq. 9-4 for the chymotrypsin-catalyzed hydrolysis of several specific ester substrates at pH 5.0 (Table 9-2). Several

TABLE 9-2 Specific Rate Constants and Equilibrium Constants Derived from Stopped Flow Measurements of Individual Steps in the Acyl-Enzyme Pathway for the Chymotrypsin-Catalyzed Hydrolysis of Specific Ester Substrates at pH 5.0 and 25°C[a]

Substrate	k_2(sec^{-1})	k_3(sec^{-1})	K_s(mM)	$\dfrac{K_s k_3}{k_2 + k_3}$ (mM)[b]	$K_{m(app)}$(mM)[c]
N-acetyl-L-Trp ethyl ester	35 ± 9	0.84	2.1 ± 0.6	0.05	0.08
N-acetyl-L-Phe ethyl ester	13 ± 2	2.2	7.3 ± 1.5	1.1	1.3
N-acetyl-L-Tyr ethyl ester	83 ± 24	3.1	18 ± 6	0.6	0.8
N-acyl-L-Leu methyl ester	3.2 ± 0.4	0.19	93 ± 11	5.2	—

[a] Data of Brandt, Himoe, and Hess[24] as reported by Hess, McConn, Ku, and McConkey.[25]

[b] Calculated from mean values as recorded. To be compared with experimental values of $K_{m(app)}$ from steady-state measurements.

[c] Experimental values from steady-state rate data with $[S]_0 \gg [E]_0$ using pH-stat titration.

important conclusions may be drawn from this work. First, it is clear that an intermediate is formed and then decomposes with these specific substrates, just as we have seen with Bernhard's furylacryloyl substrates. Second, the individual rate constants and equilibrium constants determined in these studies are consistent with the steady-state kinetic parameters k_{cat} and $K_{m(app)}$ if Eq. 9-4 is assumed. That is, within experimental error, Eqs. 9-6 and 9-7 are demonstrated experimentally to be valid. Third, for all of these esters, k_3 is much smaller than k_2, which agrees with earlier suggestions[6] that in general, for ester substrates, deacylation is rate determining. It is also, of course, consistent with the findings of Bernhard mentioned earlier concerning the rate-determining deacylation of his nonspecific ester substrates.

More recent work of Hess et al.[25] with the specific amide substrate furyl-acryloyltryptophanamide **VI** employed the temperature jump technique to measure rates of processes in the microsecond time scale, stopped-flow meth-

VI

ods for the millisecond range, and conventional spectroscopy for the steady-state liberation of ammonia. At pH 7.4 and 15°C, these experiments allowed *direct* observation of the formation of two distinct ES complexes *prior to the formation of ES' and the release of P_1* (ammonia) (Eq. 9-11). The first

$$E + S \xrightleftharpoons{K_{s1}} ES_1 \xrightleftharpoons{K_{s2}} ES_2 \xrightarrow{k_2} \overset{P_1}{\nearrow} ES' \xrightarrow{k_3} E + P_2 \qquad (9\text{-}11)$$

step, leading to ES_1, is very fast, reaching a steady state in just over 1 msec. It is reported to be pH independent. The second step is about 100 times slower than the first and is *not* pH independent. The third step is rate determining for this amide substrate with a half-time of just under 1 minute. It results in the production of ammonia and the acyl-enzyme intermediate, which then reacts in a final (unobserved) fast step to regenerate free enzyme and the final product acid. The exact nature of intermediates ES_1 and ES_2 has not yet been determined, although on the basis of arguments to be presented in Section g, it is tempting to speculate that they may be the Michaelis complex and a tetrahedral intermediate, respectively. Much work remains to be done using these fast reaction techniques to monitor changes in the physical properties of the enzyme, substrates, and inhibitors at early stages in the reaction. Parenthetically, it may be pertinent to note that findings of Bernhard and Gutfreund[24] on the trypsin-catalyzed hydrolysis of some furylacryloyl specific ester substrates are also consistent with Eq. 9-11.

e. Specificity

The general concept of enzyme specificity was introduced in Chapter 6. It will be recalled that specificity is manifested in several forms for any given enzyme (or for that matter for any given chemical reagent, although enzyme specificity is usually narrower than most reagent specificities). Thus an enzyme will exhibit substrate specificity, being more or less selective in the structural or steric variation it will tolerate in substances. There will be reaction specificity, in the case of chymotrypsin, for example, where carboxylic acid amides or esters are hydrolyzed, but phosphoric acid amides or esters are not. And there will be medium specificity defining the tolerable limits upon pH, ionic strength, solvent composition, and so on. In the present discussion, our primary interest is in the substrate specificity of chymotrypsin, as reflected in the relative magnitudes of kinetic parameters governing hydrolysis rates as substrate structure and stereochemistry are varied.

Having established that chymotrypsin catalysis is at least a three-step process (binding, acylation, and deacylation), we must realize that elements of substrate specificity may be manifest in each of these steps and in fact the specificity requirements of the various steps may indeed be independent of one another. Ideally, then, we should consider the effects of substrate structure variation on each of the independent kinetic parameters K_s (binding), k_2 (acylation), and k_3 (deacylation). Unfortunately, values for these independent parameters are not presently available for any but a few substrates, so that for the moment we shall have to limit our discussion to those conclusions that can be derived from the wealth of steady-state data now available. Bender and Kezdy[6] have considered this problem in some detail and concluded that in general the only steady-state parameter that has unequivocal significance in terms of the acyl-enzyme pathway is the ratio $k_{cat}/K_{m(app)}$, which always is equal to k_2/K_s (see Chapter 5). Furthermore, in cases where it can be demonstrated that for all substrates being compared $k_2 \gg k_3$, then k_{cat} is a direct measure of the deacylation step (k_3) and can be used in meaningful comparisons. Finally, it may be to some extent valid to use K_i data for competitive inhibitors which are substrate analogs to gain some appreciation for the effects of structural changes on binding specificity.

Although in principle it should be possible to "feel" our way to a reasonably accurate spacial picture of the chymotrypsin-active site solely on the basis of what suitable kinetic parameters report about the relative "correctness" of various structural features of substrates and inhibitors, our job is considerably simplified by the current availability of X-ray crystallographic data on the structure of chymotrypsin.[27] Hence we shall draw on this direct structural information freely in our discussion, even though historically many correct conclusions concerning the structure of the chymotrypsin-active site were arrived at on the basis of kinetic data alone.

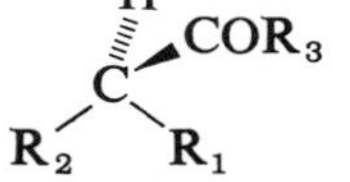

VII

Most chymotrypsin substrates may be represented by the general formula **VII**, drawn here with the L-configuration. It was recognized early[28] that R_2 is preferably one of the aromatic amino acid side chains (i.e., the benzyl group of phenylalanine, the p-hydroxybenzyl group of tyrosine, or the indolylmethylene group of tryptophan) and that **VII** should have the L-configuration. In specific substrates (defined earlier in Section c) R_1 is an acylated α-amino group R'CONH— and R_3 is the incipient amine or alcohol leaving group —NHR or —OR.

With reference to this general formula, most discussions of chymotrypsin specificity assume that when a specific substrate is productively bound at the active site of the enzyme, each of the four groups about the asymmetric carbon atom must correctly occupy a corresponding defined position or subsite within the active site. Cohen[29] refers to these subsites as *am* for the acylamino group R_1, *ar* for the aromatic side chain R_2, *n* for the scissile amide or ester group R_3, and *h* for the α-hydrogen. Niemann[30] refers to the same set of subsites as ρ_1, ρ_2, ρ_3, and ρ_4, respectively. Conceivably, each of the four subsites might interact with its corresponding group through specific binding forces; but if the active site of chymotrypsin is essentially rigid, only two such specific binding interactions are required to define the positions of all four groups of a particular configurational isomer within the active site and thus give rise to the observed steric (or configurational) specificity of the enzyme.

It has been established beyond reasonable doubt that subsites *ar*, *am*, and *n* do in fact exist at the chymotrypsin active site. The *ar* subsite, which shows up very clearly in electron density maps from X-ray diffraction data,[31] is a hydrophobic pocket whose dimensions are estimated to be 10–12 Å by 5.5–6.5 Å by 3.5–4.0 Å and into which the R_2 side chain of specific chymotrypsin substrates may comfortably fit. Through hydrophobic forces,* this *ar* pocket provides the primary binding interaction between the enzyme and its most effective substrates and competitive inhibitors.† The *am* subsite interacts by hydrogen

---

* The free energy of hydrophobic bonding that is realized when an aromatic group R_2 leaves its aqueous environment and occupies the *ar* pocket of the chymotrypsin-active site is closely related to the free energy associated with the transfer of an aromatic hydrocarbon from water solution to some nonaqueous phase such as ether or benzene.[32] The magnitude of this free energy for the transfer of benzene from water to benzene is about −5 kcal/mole. If all of this driving force were available for the transfer of a phenyl group on a chymotrypsin substrate from water to the hydrophobic *ar* pocket, and if this were the *only* binding force responsible for the association of the enzyme-substrate complex, a K_s of about $2.5 \times 10^{-4} M$ for that substrate would result. The actual K_s for acetyl-L-phenylalanine ethyl ester is about $2 \times 10^{-3} M$, which is obviously well within the hypothetical limit we have assumed could be accounted for. This one example lends credence to the statement of Jencks[33] that "'hydrophobic forces' are probably the most important single factor providing the driving force for noncovalent intermolecular interactions in aqueous solution," where he defines hydrophobic bonding as "an interaction of molecules with each other which is stronger than the interaction of the separate molecules with water and which cannot be accounted for by covalent, electrostatic, hydrogen bond, or charge transfer forces."

† These include not only substrate analogs such as N-acylated aromatic amino acids (both *L*- and *D*- enantiomers) and the *D*-enantiomers of specific *L*-substrates, but also simpler aromatic compounds such as indole, β-phenylpropionic acid, and proflavine.

bonding with the amido —NH— of R_1 in specific substrates. The hydrogen-bond acceptor of this subsite has been tentatively identified by the crystallographers[31] as the peptide carbonyl group of Ser 214. Although the free energy associated with this hydrogen bond is comparable to that associated with hydrophobic bonding in the *ar* pocket, it apparently does not contribute to the overall binding interaction between specific substrates and the enzyme (i.e., to K_s) but rather contributes directly to catalysis (to k_2 and k_3). We shall have more to say about this later.

The catalytic subsite *n* contains (among other residues essential to catalysis) the active Ser 195 residue, whose implication in the acyl-enzyme pathway by which chymotrypsin operates was discussed in the previous section. It must be correctly occupied by the scissile amide or ester bond if catalysis is to occur, but for most substrates* no significant binding interactions have been attributed to subsite *n* prior to the formation of a covalent bond to Ser 195 upon acylation.

Cohen[29] has proposed that the active site of chymotrypsin includes a restricted site (subsite *h*) only large enough to accommodate the α-hydrogen of substrates with formula **VII**. He based this proposal primarily upon (1) the essentially complete failure of chymotrypsin to catalyze the hydrolysis of compounds where R_1, R_2, and R_3 are optimal but the α-hydrogen is replaced by a more bulky group such as methyl, and (2) the L-specificity of the enzyme toward compounds where R_1 has no hydrogen-bond donor and thus should show no particular preference for the *am* subsite over the presumed *h* subsite if the latter were not sterically restrictive. The crystallographic results indicate no such restrictive *h* subsite.[27] However, they do show that when formyl-L-tryptophan (the product of an excellent specific substrate) is bound at the active site in crystalline chymotrypsin, it assumes a conformation such that the α-hydrogen is in contact with the carbonyl oxygen of the formyl group.[31] This might imply that intramolecular steric hindrance between an α-substituent other than hydrogen and the carbonyl oxygen of the

---

* Recently Cohen and his co-workers investigated some substrates that may experience a specific noncovalent binding interaction at or near subsite *n*.[34] Thus, for example, for substrates of the type

$$\text{L-C}_6\text{H}_5\text{CH}_2\overset{\displaystyle |}{\underset{\displaystyle R_1}{\text{CH}}}\text{COR}_3$$

if R_3 is —OCH_3, changing R_1 from CH_3CONH— to CH_3CO_2— brings about a reduction in $k_{cat}/K_{m(app)}$ of over two orders of magnitude as expected due to the hydrogen bonding requirements of subsite *am*. But if R_1 is CH_3CO_2—, changing R_3 from —OCH_3 to —$OCH\ CON\ H$ results in a surprising *increase* in $k_{cat}/K_{m(app)}$ of over two orders of magnitude. This could mean that loss of the hydrogen-bonding interaction at *am* can, with properly constituted substrates, be compensated for by a similar interaction near subsite *n*.

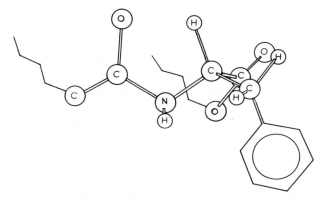

FIGURE 9-2. A specific substrate of chymotrypsin drawn to indicate the probable conformation it assumes when bound productively at the enzyme active site. The drawing is patterned after the conformation of formyl-L-tryptophan bound to chymotrypsin in the crystalline state, as determined by X-ray diffraction studies.[27]

α-N-acyl group would prevent a compound from taking up the "correct" conformation critical to proper alignment of the scissile bond at subsite *n*. Figure 9-2 is drawn to indicate the probable orientation of atoms in that "correct" conformation.

Table 9-3, 9-4, and 9-5 provide quantitative data concerning the side chain specificity and stereospecificity of chymotrypsin. Table 9-3 clearly shows the

TABLE 9-3 Kinetic Constants for the Chymotrypsin-Catalyzed Hydrolysis of Some α-N-Acetyl-L-amino Acid Methyl Esters[a]

Substrate L-AcNHCHRCO$_2$CH$_3$	$k_{cat}/K_{m(app)}$ (sec^{-1} M^{-1})
R = H	0.42
R = CH$_3$	1.7
R = C$_2$H$_5$	20
R = n-C$_3$H$_7$	270
R = n-C$_4$H$_9$	1,250
R = n-C$_5$H$_{11}$	8,200
R = n-C$_6$H$_{13}$	2,100
R = CH(CH$_3$)$_2$	1.4
R = CH$_2$CH(CH$_3$)$_2$	1,300
R = CH$_2$C$_6$H$_5$	42,000
R = CH$_2$C$_6$H$_{11}$	80,000

[a] In aqueous solution at 25.0°, pH 7.90, and 0.01M in NaCl. Adapted from J. B. Jones, T. Kunitake, C. Niemann, and G. Hein, *J. Am. Chem. Soc.*, 87, 1777 (1965).

improvement in $k_{cat}/K_{m(app)}$ which results from improved hydrophobic inter-action between the hydrocarbon side chain R_2 and the *ar* pocket as the size of the side chain increases. Several particular features of the table are to be noted. First, it is clear that a cyclic structure makes it possible to fit more carbons in the pocket than is the case with an open chain. (Compare, for example, the benzyl and hexahydrobenzyl side chains, which fit, to the *n*-hexyl side chain, which apparently does not.) Second, aromaticity in the side chain is of no particular advantage. Finally, branching at the β-carbon is undesirable. Thus the isopropyl side chain is far worse than the *n*-propyl side chain, whereas isobutyl and *n*-butyl side chains are comparable.

Table 9-4 also deals with variation in R_2 structure, and again we see the effect of hydrocarbon side chain size on $k_{cat}/K_{m(app)}$ within the L-series. However, a comparison of the entries in the table for $k_{cat}/K_{m(app)}$ (k_2/K_s)

TABLE 9-4 Kinetic Constants for the Chymotrypsin-Catalyzed Hydrolysis of Some α-N-Acetylamino Acid *p*-Nitrophenyl Esters[a]

Substrate, AcNHCHR₂CO₂C₆H₄NO₂	Config- uration	$k_{cat}/K_{m(app)}$ $(M^{-1}sec^{-1})$	k_3 norm[b]	$\dfrac{k_3{}^L}{k_3{}^D}$	K_i for Corre- sponding D amide (mM)
$R_2 = CH_2$—indole	L	10,600,000	560	18,000	1.48
	D	—	0.0302		
$R_2 = CH_2$—C_6H_5	L	19,600,000	296	6,400	10.6
	D	—	0.047		
$R_2 = CH_2CH(CH_3)_2$	L	827,000	24.1	148	141
	D	—	0.166		
$R_2 = H$	—		5,840	1.0	1.0	692

[a] In 1.6% (V/V) acetonitrile at 25°, pH 7.0, ionic strength 0.1M, phosphate buffer. Adapted from data of D. W. Ingles and J. R. Knowles, *Biochem. J.*, 104, 369 (1967); 108, 561 (1968).

[b] Relative values of specific deacylation rate constants normalized for inherent differences in reactivity due to steric and electronic factors unrelated to enzyme catalysis. See Ingles and Knowles (1967).

with those for the deacylation rate constant k_3 and the inhibition constant K_i indicates that the interaction at subsite *ar* is important in catalysis as well as in binding. Of particular interest in this regard are the "stereospecificity ratios" $k_3{}^L/k_3{}^D$.[35] These imply that as the interaction potential at subsite *ar* gets weaker, it becomes increasingly possible (although still only slightly so) to realize catalysis for D-enantiomers, presumably by interchanging the contents of subsite *ar* with those of subsite *am* (or of subsite *h*, although this is less likely in view of our previous discussion of the nature of this "subsite").

TABLE 9-5 Kinetic Constants for the Chymotrypsin-Catalyzed Hydrolysis of Some α-Substituted β-Phenylpropionate p-Nitrophenyl Esters[a]

Substrate, $C_6H_5CH_2CHR_1CO_2C_6H_4NO_2$	Config-uration	$k_{cat}/K_{m(app)}$ $(M^{-1} sec^{-1})$	k_3 (sec^{-1})	$\dfrac{k_3{}^L}{k_3{}^D}$	K_i for corres-ponding D amide (mM)
$R_1 = CH_3CONH-$	L	19,600,000	94.8	6,400	10.6
	D	—	0.0148		
$R_1 = CH_3COO-$	L	1,510,000	9.2	9	8.4
	D	—	1.03		
$R_1 = H-$	—		0.074	1	4.55

[a] In 1.6% (v/v) acetonitrile at 25°, pH 7.0, ionic strength 0.1M, phosphate buffer. Adapted from data of D. W. Ingles and J. R. Knowles, *Biochem. J.*, **108**, 561 (1968).

Table 9-5 is concerned with the requirements for a hydrogen-bond donor at subsite *am*. Again we see decreases in all the kinetic parameters when the hydrogen-bonding capability of the R_1 side chain is removed, or (except for K_i) when the D-isomer replaces the L-isomer. But note that the apparent overall binding energy (as reflected in K_i values) is not markedly affected by structural changes in R_1, whereas deacylation (k_3) and acylation (k_2/K_s) are, and stereospecificity ($k_3{}^L/k_3{}^D$) is very strongly affected. These data provide a basis for the claim made earlier that hydrogen bonding at subsite *am* contributes primarily to specificity as reflected in catalysis rather than specificity as reflected in binding.

The explanation for this may be related to the following:

1. Formation of a second attachment of the substrate to the enzyme-active site at subsite *am* freezes the substrate in position with the scissile bond of R_3 correctly positioned at subsite *n*. But this positioning results in losses of rotational degrees of freedom for the substrate and gives rise to an unfavorable entropic contribution to the free energy of binding, which tends to cancel out the energy of the hydrogen bond.[36]

2. The full energy of the hydrogen bond at *am* cannot in fact be realized in net binding energy since when it is formed, water molecules that were associated with the R_1 group must be displaced.[34]

3. To form the hydrogen bond at *am*, the substrate must assume an unfavorable high-energy conformation, which is paid for by the energy of the hydrogen bond.[35]*

What is important to realize here is that the freezing of the substrate in

---

* NMR studies[37] tend to refute this, indicating that tryptophan has a similar conformation in solution and when bound to chymotrypsin.

proper position, the desolvation of the substrate, and perhaps the conformational strain on the substrate as well, *all contribute directly to rapid catalysis.* Thus the energy of the hydrogen bond is expended specifically to reduce or remove free-energy barriers that would otherwise contribute directly to the activation energies of acylation and deacylation.

f. Identity of Residues at the Active Site

The involvement of the hydroxyl group of Ser 195 in the acylation and deacylation steps of chymotrypsin-catalyzed hydrolyses has already been discussed. We have also noted the possible role of the carbonyl oxygen of Ser 214 as a hydrogen-bond acceptor at subsite *am* in the chymotrypsin active site. Other amino acid residues implicated as functionally important constituents of the chymotrypsin active site include[38] His 57, Asp 102, Ile 16, and Asp 194.

The direct involvement of at least one of chymotrypsin's two histidine residues (His 40 and His 57) in the catalytic action of the enzyme was first suspected[39] on the basis of the pH dependence of catalysis. Toward any neutral substrate, it can be observed that the activity of chymotrypsin (as measured by $k_{cat}/K_{m(app)}$) is minimal below pH 6, rises to a maximum at about pH 8, and falls again to a low level above pH 10. As discussed in Chapter 5, such bell-shaped pH-activity profiles are most simply interpreted to mean that two ionizing groups of the enzyme (a basic group that must be unprotonated and an acidic group that must be protonated) are involved in catalysis. For chymotrypsin, appropriate analysis of the pH-profile data indicates that these two presumed groups have apparent pK_a values of about 7 and about 8.5, respectively. As inspection of Table 5-2 will show the imidazole side chain of a histidine residue is the prime candidate to account for the pH dependence of chymotrypsin at low pH. Recent studies[25] on the pH dependence of the individual specific kinetic constants k_2 and k_3 of Eq. 9-4 establish that both acylation and deacylation (but not the initial binding step) require that this presumed histidine residue be in its unprotonated neutral state.

By the affinity-labeling technique (see Chapter 3) Shaw and co-workers[40] demonstrated the importance of His 57 in particular and established that it was present at the chymotrypsin active site. These workers synthesized L-1-tosylamido-2-phenylethylchloromethyl ketone (TPCK) **VIII**, a structural analog of specific substrate derivatives of L-phenylalanine such as **IX**. When

$$\underset{\textbf{VIII}}{C_6H_5CH_2-\overset{\overset{H}{|}}{\underset{\underset{NHSO_2C_7H_7}{|}}{C}}\text{-}COCH_2Cl} \qquad \underset{\textbf{IX}}{C_6H_5CH_2-\overset{\overset{H}{|}}{\underset{\underset{NHSO_2C_7H_7}{|}}{C}}\text{-}CO_2CH_3}$$

chymotrypsin was incubated with ^{14}C-labeled TPCK, an alkylation reaction resulted in which the stoichiometric incorporation of one equivalent of radioactive label per mole of enzyme was paralleled by complete enzyme inactivation. Both the label incorporation and the concomitant loss of enzyme activity were found to be retarded by the presence of β-phenylpropionic acid (a competitive inhibitor) and prevented by prior DFP inactivation or urea denaturation of the enzyme.

Examinations of partial and complete amino acid hydrolysates from TPCK-inactivated chymotrypsin revealed that alkylation was directed exclusively toward $N^{\varepsilon 2}$ of the imidazole ring of His 57. Moreover, it was later established[41] that as a substrate analog, TPCK does indeed form a reversible, competitive inhibitor-type complex with chymotrypsin prior to alkylation ($K_i = 3 \times 10^{-4}M$) and that[42] the rate of alkylation by TPCK is subject to the same pH dependence that characterizes chymotrypsin-catalyzed hydrolyses [$pK_{1(app)} = 6.8$; $pK_{2(app)} = 8.9$]. A number of other affinity-labeling reagents for the active site of chymotrypsin have been investigated,[43,44] among which several behave much like TPCK implicating His 57.

The crystallographic studies of Blow and co-workers[27] show that in both native chymotrypsin crystals and crystalline complexes of the enzyme with hydrolysis products or inhibitors, the hydroxyl group of Ser 195 and the $N^{\varepsilon 2}$ of His 57 assume positions within hydrogen-bonding distance of each other. Furthermore, the side chain of Asp 102 is buried in a hydrophobic environment beneath the side chain of His 57 with its carboxyl group in close contact with the $N^{\delta 1}$ of the imidazole ring. The close spacial proximity of these three groups among which polar interactions of various sorts (e.g., hydrogen bonding, ion pairing, dipole-dipole, or charge-dipole) may be possible is undoubtedly significant in terms of the mechanism of catalysis, which we will consider in the next section.

The acidic group with an apparent pK_a of 8.5 which is responsible for the falloff of chymotrypsin activity at high pH has been conclusively identified by Labouesse, Oppenheimer, and Hess[45,46] as the protonated α-amino group of Ile 16. Realizing that the one critical step leading to the conversion of inactive chymotrypsinogen into an active enzyme (π-chymotrypsin) is the cleavage of the peptide bond between Arg 15 and Ile 16 (see Section a of this chapter), these investigators were also able to elucidate essential relationships between the inactivity of chymotrypsin at high pH and the inactivity of the zymogen. This work and other findings that relate to it are discussed in some detail in recent reviews.[25,38] In what follows we will consider some of the primary lines of evidence and major conclusions.

If chymotrypsinogen is treated with acetic anhydride, a number of lysine ε-amino groups and the α-amino N-terminus at Cys-1 are acetylated. Nevertheless, upon tryptic activation this acetylated chymotrypsinogen gives rise to an acetylated δ-chymotrypsin whose activity is indistinguishable from that of δ-chymotrypsin itself. However, if this acetylated δ-chymotrypsin is then reacetylated with ^{14}C-labeled acetic anhydride, one observes closely parallel loss of enzymatic activity and incorporation of radioactive label specifically on the α-amino group of Ile 16. These results in themselves establish that the free α-amino group of Ile 16 is essential to chymotrypsin activity.

Acetylated chymotrypsinogen and acetylated (active) δ-chymotrypsin differ by only one titratable acidic group, the α-ammonium group of Ile 16. The same is true for reacetylated (inactive) δ-chymotrypsin *versus* acetylated (active) δ-chymotrypsin. Difference titration data for these two pairs of proteins show that the α-ammonium group of Ile 16 has a pK_a of about 8.5, which of course is consistent with the hypothesis that this group is responsible for the drop in chymotrypsin activity above pH 8. Incidentally, that drop in activity above pH 8 is the result of an unfavorable effect on K_s alone.[47] The catalytic rate constants k_2 and k_3 are unaffected by the ionization state of Ile 16.

The activation of chymotrypsinogen is accompanied by a conformational change in the protein as evidenced by large differences in the circular dichroism spectra of the zymogen and active δ-chymotrypsin or α-chymotrypsin at neutral pH. (The α- and δ-forms have very similar spectra.) This conformational change is presumably responsible for establishing the active site of the enzyme, giving rise to activity. Now the circular dichroism spectra of chymotrypsinogen, acetylated chymotrypsinogen, and reacetylated (inactive) δ-chymotrypsin are similar and are pH independent. However, the circular dichroism spectra of acetylated (active) δ-chymotrypsin change as a function of pH. As the pH is increased from 8 to 10, the CD spectra of the active enzyme resemble increasingly the spectra of the inactive zymogen, and this pH dependency is just the same as the pH dependency of K_s for the action of δ-chymotrypsin on the specific neutral substrate, N-acetyl-L-tryptophanamide.

All this is interpreted to mean that chymotrypsin can assume two stable conformational states, an active conformation which is favored when the α-amino group of Ile 16 is protonated and thus bears a positive charge, and an inactive conformation which predominates when the α-amino group of Ile 16 is electrically neutral (as in the zymogen or when acetylated or at high pH). The difference in activity between these two conformations is a result of the relative inability of the high pH inactive conformation to bind substrate productively at the active site.

These conclusions have in large part been substantiated by the crystallographic structural studies on α-chymotrypsin[27] and on chymotrypsinogen.[48] In α-chymotrypsin crystallized at pH 4.2 (the low pH active conformation), it is apparent that the positively charged α-ammonium group of Ile 16 and the negatively charged carboxylic side chain of Asp 194 form a tight ion pair in a hydrophobic environment toward the interior of the enzyme and away from the substrate binding site. On the other hand, in chymotrypsinogen (whose conformation is presumably quite similar to that of the high pH inactive form of chymotrypsin) Asp 194 forms an internal ion pair instead with His 40. As a result, a segment of the peptide chain which includes Asp 194 partially occupies the substrate binding site of the enzyme.

g. A Possible Mechanism of Action

Progress in the investigation of the many aspects of chymotrypsin catalysis and structure since 1950 has been marked along the way by many attempts to summarize all the known facts in terms of a more or less comprehensive mechanism of action for chymotrypsin action. A number of these proposals from the earlier literature have been summarized by Cunningham.[7] The mechanism schematically depicted in Figure 9-3 and discussed next is based primarily upon recent arguments presented by Wang,[49] Blow et al.,[50] Caplow,[51] Polgar and Bender,[52] and Fersht.[53,54]

In Figure 9-3 the simple acyl-enzyme pathway of Eq. 9-4 has been expanded to include two additional intermediates, a tetrahedral intermediate (ES_{tet}) directly preceding the formation of the acyl-enzyme and P_1, and a second tetrahedral intermediate (ES'_{tet}) preceding the regeneration of free enzyme and P_2. This is justified on several grounds. First, both acylation and deacylation may be clearly categorized as examples of nucleophilic substitution at a carbonyl carbon. It is an accepted fact that most such reactions proceed through a tetrahedral intermediate as shown for the general case in Eq. 9-12 where $Y:^-$ is a nucleophile and $X:^-$ is an appropriate leaving group.[55] There is no reason to believe that such tetrahedral intermediates would not be formed in reactions of this type involving enzymes.

$$\begin{array}{c} R \\ \diagdown \\ \\ X \diagup \end{array}\!\!C\!=\!\ddot{O}\!:\ +\ Y\!:^- \underset{k_{-1}}{\overset{k_1}{\rightleftharpoons}} \begin{array}{c} R_{\prime\prime\prime} \\ \ \\ X \diagup \end{array}\!\!\overset{\overset{Y}{|}}{C}\!-\!\ddot{O}\!:^- \ \overset{k_2}{\longrightarrow}\ \begin{array}{c} R_{\prime\prime\prime} \\ \ \\ Y \diagup \end{array}\!\!C\!=\!O\ +\ X\!:^- \tag{9-12}$$

Second, it has been observed[53,54] that the pH dependence of the kinetics of chymotrypsin-catalyzed hydrolyses of certain anilide and hydrazide substrates requires the existence of an additional intermediate between the Michaelis complex and the acyl enzyme. The actual observation is that for these substrates, k_{cat} varies with pH according to an ionization curve that cannot be attributed to the real ionization of any acid or base in the enzyme-substrate complex. The apparent anomalous pK_a associated with this curve has been attributed to a pH-dependent change in the rate-determining step for acyl enzyme formation. At low pH, the rate-determining step in acylation is the formation of an intermediate from the Michaelis complex, whereas at high pH, the breakdown of this intermediate to the acyl enzyme is rate limiting. Although the kinetics yield no information concerning the nature of this intermediate, it is reasonable to suspect that it is indeed the tetrahedral adduct of Ser 195 with the amide carbonyl of the substrate as indicated in Figure 9-3.

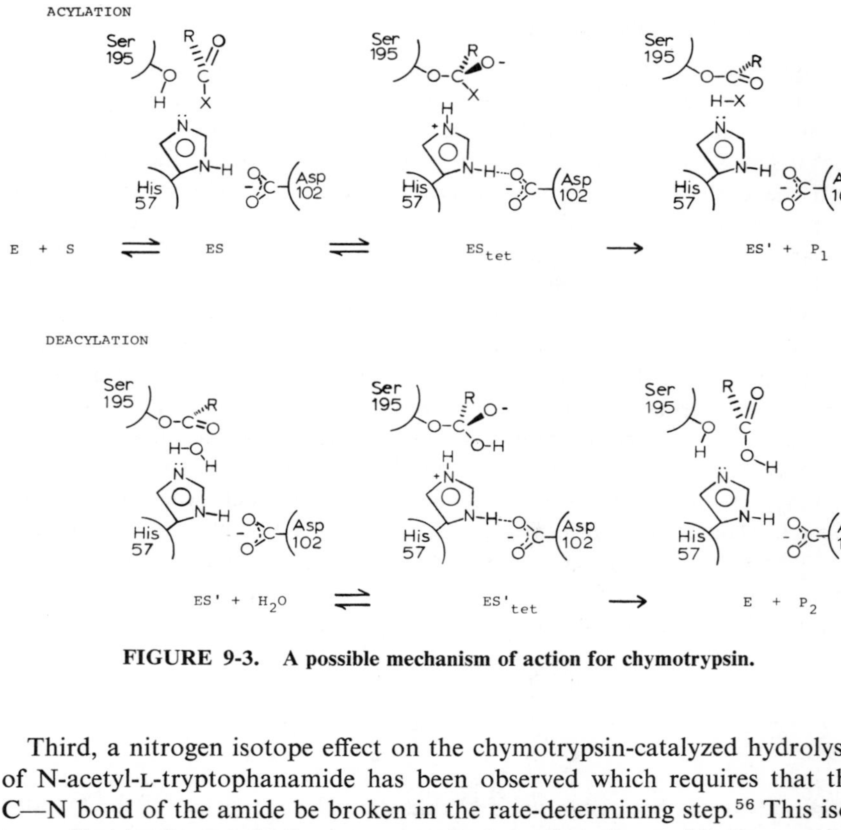

FIGURE 9-3. A possible mechanism of action for chymotrypsin.

Third, a nitrogen isotope effect on the chymotrypsin-catalyzed hydrolysis of N-acetyl-L-tryptophanamide has been observed which requires that the C—N bond of the amide be broken in the rate-determining step.[56] This isotope effect is of precisely the same magnitude as that observed in the reaction of amides with hydroxide ion, a reaction known to proceed through a tetrahedral intermediate. In view of the evidence cited in the previous paragraph, it would be most interesting to see if a similar nitrogen isotope effect is found at high pH with substrates that give rise to an anomalous dependence of k_{cat} on pH. If so, the isotope effect should disappear at low pH if the interpretation of the pH effect on k_{cat} is correct as stated.

Fourth,[51] the rates of acylation of chymotrypsin by substituted phenyl acetates increase as a function of electron withdrawal by substituents on the benzene ring of the leaving group. On the other hand, acylation rates with substituted specific anilide substrates of chymotrypsin *decrease* as a function of electron withdrawal in the leaving group. If the mechanism of acylation is the same in both cases (and there is no compelling reason to believe otherwise), these observations require that mechanism to have at least two steps which are oppositely affected by electron withdrawal. In the two-step tetrahedral mechanism of Eq. 9-12 (where Y is the nucleophilic Ser 195 of chymotrypsin) the attack step will be facilitated by electron withdrawal in X, which tends to enhance the electropositive character of the carbonyl carbon. On the other hand, the second step in which the leaving group X is expelled is discouraged by electron withdrawal in X. The reason for this is not apparent from Eq. 9-12 as it is written, but when X is a poor leaving group (as an anilide ion $C_6H_5NH^-$ would be), protonation of the leaving group must accompany C—X bond cleavage as in fact it does in Figure 9-3. Electron-withdrawing substituents in X make this protonation more difficult, as reflected in decreased values of k_2. It will now be apparent that the seemingly anomalous substituent effects upon acylation rates can be accounted for in terms of Eq. 9-12 where k_1 is rate determining for the substituted phenyl acetates and k_2 is rate determining for substituted anilide substrates.

Finally, in support of expanding Eq. 9-4 to include more intermediates we have the direct evidence of Hess et al.[25] cited earlier in Section d concerning the chymotrypsin-catalyzed hydrolysis of the specific amide substrate furylacryloyltryptophanamide VI.

It should be noted that although the evidence cited applies specifically only to the acylation steps of chymotrypsin-catalyzed hydrolysis, it is generally accepted that the mechanism of deacylation must be the "microscopic reverse" of acylation.[6] That is, deacylation must proceed through the same sequence of intermediates in reverse order with water replacing the leaving group HX as shown in Figure 9-3. Thus whatever intermediates are found experimentally to be associated with acylation may be assumed to be included in deacylation as well, and vice versa.

As demanded by the pH-dependency evidence, the chemical modification evidence and the X-ray crystallographic evidence cited in Section f, Figure 9-3 proposes a specific role in acylation and deacylation for the interacting constellation of amino acid side chains on Ser 195 His 57 and Asp 102. Upon binding at the active site, the substrate is frozen in position with its carbonyl carbon in contact with the hydroxyl oxygen of Ser 195 and the proton acceptor atom of its leaving group X in such a position that the proton of the serine hydroxyl group can be transferred to it by the intervening $N^{\varepsilon 2}$ nitrogen atom of His 57. The relative positioning of these five atoms directly involved in the bond-making and bond-breaking processes of acylation is critical, making a large (though quantitatively undefinable) contribution to the specific rate enhancements associated with chymotrypsin catalysis.*

The formation of the tetrahedral intermediate by attack of the serine oxygen on the carbonyl carbon is accompanied by transfer of the hydroxyl proton to the $N^{\varepsilon 2}$ of His 57 in a general-base catalyzed bond-making step. For ester substrates, this step controls the rate of acylation. The ensuing bond-breaking step leading to the acyl-enzyme intermediate and an alcohol or amine is general-acid catalyzed by the imidazolium ion formed in the first step. This second step is normally rate controlling for amide substrates, which for them allows the buildup of an observable concentration of the tetrahedral intermediate,† possibly accounting for the findings of Hess et al. cited in Section d.

The possible role of Asp 102 in this process is difficult to assess with confidence. Polgar and Bender[52] suggest that by hydrogen bonding to $N^{\delta 1}$ of His 57 it serves merely to stabilize the imidazolium ion formed along with the tetrahedral intermediate against premature loss of its proton. (This is the role illustrated in Figure 9-3.) They advance strong arguments against the hypothesis of Blow et al.[50] that a hydrogen-bonded "charge relay system" involving Asp 102, His 57, and Ser 195 exists and accounts for the high apparent nucleophilicity of the serine hydroxyl group (Figure 9-4). The reader is referred to the original literature for a full discussion on this point.

The deacylation steps of Figure 9-3 are just the acylation steps in reverse where water replaces the amine or alcohol molecule HX. Here, since the substrate which water attacks is always the acyl enzyme (an ester), the attack

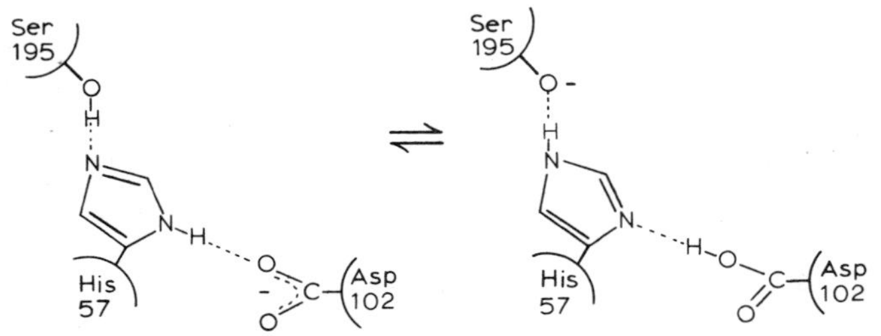

FIGURE 9-4. The hydrogen-bonded "charge-relay system" postulated by chymotrypsin crystallographers[50] to account for the enhanced nucleophilicity of Ser 195 at the active site of chymotrypsin.

---

* See Chapter 7 on Approximation of Reactants and Orientation Effects.

† The work of Fersht[53,54] cited earlier indicates that breakdown of the tetrahedral intermediate may not be rate controlling for some amide substrates at low pH. Fersht also disputes that the putative tetrahedral intermediate accumulates significantly, even when its breakdown is rate limiting.

step will presumably always be rate limiting and direct observation of tetrahedral intermediates in deacylation should not be possible.

The mechanism for chymotrypsin action shown in Figure 9-3 may not be regarded as definitely established, except in broadest general outline. The role of hydrogen bonding between amino acid side chains in the active site and between them and the substrate has been portrayed differently by various authors (see, in particular, ref. 49 and ref. 50). Certain differences between the behavior of ester and amide substrates[38] are not well understood and might not be consistent with this mechanism, which is presented as applicable to both. Although the rigid active-site hypothesis assumed in Figure 9-3 seems consistent with the known facts, it is possible that significant protein conformational changes of functional significance ("induced-fit" hypothesis) may be involved,[57] perhaps giving rise to additional intermediate enzyme-substrate complexes. It is to be hoped that the continued imaginative application of techniques such as X-ray crystallography, stopped flow and relaxation kinetic studies, isotope effects, and various kinds of spectroscopy will soon clear up these and other uncertainties concerning the action of chymotrypsin.

Carboxypeptidases A (EC 3.4.3.1)

a. Structure and Function

Studies on the structure and function of bovine pancreatic carboxypeptidase A (CPA)* have been under way since 1937 when the enzyme was first crystallized by Anson.[58] Like chymotrypsin, CPA originates in the form of an inactive zymogen (procarboxypeptidase A), which is activated upon exposure to the protease trypsin. *In vivo*, this occurs in the small intestine. The zymogen itself, as extracted from an acetone powder of beef pancreas glands, has been found to be an oligomer of two or three protein subunits, only one of which is the direct precursor of the active enzyme.[59] Reasons for the inactivity of procarboxypeptidase A and the events of tryptic activation leading to active CPA are as yet incompletely understood, but several active forms of the enzyme have been isolated. The primary structures of these forms differ only in the number of amino acid residues left behind at the N-terminus of CPA when the "activation peptide" of some 60 amino acid residues is cleaved away from the precursor. The forms are designated by Greek letters as indicated in X. The predominant form depends upon the method of

---

* Beef pancreas is not the only source of carboxypeptidase A. Other mammals and even fish give rise to similar enzymes, but the bovine enzyme has been the most widely studied and is the most readily available. All references to carboxypeptidase A in this book are to bovine pancreatic CPA in particular.

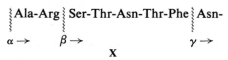

isolation.[59] CPA$_\gamma$, which is obtained by tryptic activation of a crude exudate from thawing frozen pancreas glands followed by fractional precipitations and crystallizations according to the method of Anson,[58] has been most widely studied and is commercially available from many sources. The recently determined[60] total amino acid sequence of 307 residues in a single polypeptide chain for CPA$_\alpha$ is given in Table 9-6.

Like chymotrypsin, CPA catalyzes the hydrolysis of peptide and ester linkages in a variety of substrates which vary from naturally occurring proteins and polypeptides to small synthetic model compounds. However, whereas chymotrypsin is an *endopeptidase* (it is active toward appropriate linkages regardless of their positions with respect to the ends of a peptide chain), CPA is an *exopeptidase* (its action is limited specifically to the peptide or ester linkage connecting an appropriate carboxyl-terminal residue to the remainder of the substrate molecule). Compounds that have been found to be CPA substrates may be grouped in five categories as illustrated in Table 9-7. The position of hydrolytic cleavage is the same in each, as indicated by the broken arrows through the scissile bonds.

The substrate specificity of CPA has been widely investigated, at least for peptide substrates, and reviews of this literature are available.[61-63] We note here the major features of CPA substrate specificity, because they have a definite bearing on mechanistic hypotheses we will consider later. All of the substrate types of Table 9-7 have the same kind of leaving group, a mono-substituted α-amino or α-hydroxy acid in its free ionized form. The free ionized carboxyl group and the α-position of the —NH— or —O— of the scissile bond with respect to that carboxyl group are absolute requirements (no known exceptions). Moreover, in peptide substrates the hydrogen atom of this scissile —NH— may not be replaced by any other atom or group. The side chain of the C-terminal residue must not bear a charge, and in the best CPA substrates this side chain is a large hydrophobic group (aromatic or branched aliphatic). Almost without exception, this terminal residue must have the L-configuration.*

As indicated by Table 9-7, the specificity of CPA with respect to the acyl portion of the substrate molecule is not closely defined. It may vary from a

---

* The exceptions involve some alanine polypeptides in which D-alanine was hydrolyzed from the LLLD, DLLD, LLLLD, DLLLD, LLLLLD, and DLLLLD compounds by CPA. The rate of hydrolysis of these "forbidden" bonds was two to three orders of magnitude slower than the rate of hydrolysis of C-terminal L-alanine from the corresponding LLLL, DLLL, LLLLL, DLLLL, LLLLLL, and DLLLLL isomers.[64]

TABLE 9-6 Amino Acid Sequence of CPA$_\alpha$ as Determined by Bradshaw et al.[60]

Ala	Arg	Ser	Thr	5 Asn	Thr	Phe	Asn	Tyr	10 Ala	Thr	Tyr	His	Thr	15 Leu	Asp	Glu	Ile	Tyr	20 Asp
Phe	Met	Asp	Leu	25 Leu	Val	Ala	Gln	His	30 Pro	Slu	Leu	Val	Ser	35 Lys	Leu	Gln	Ile	Gly	40 Arg
Ser	Tyr	Glu	Gly	45 Arg	Pro	Ile	Tyr	Val	50 Leu	Lys	Phe	Ser	Thr	55 Gly	Gly	Ser	Asn	Arg	60 Pro
Ala	Ile	Trp	Ile	65 Asp	Leu	Gly	Ile	His	70 Ser	Arg	Glu	Trp	Ile	75 Thr	Gln	Ala	Thr	Gly	80 Val
Trp	Phe	Ala	Lys	85 Lys	Phe	Thr	Glu	Asn	90 Tyr	Gly	Gln	Asn	Pro	95 Ser	Phe	Thr	Ala	Ile	100 Leu
Asp	Ser	Met	Asp	105 Ile	Phe	Leu	Glu	Ile	110 Val	Thr	Asn	Pro	Asn	115 Gly	Phe	Ala	Phe	Thp	120 His
Ser	Glu	Asn	Arg	125 Leu	Trp	Arg	Lys	Thr	130 Arg	Ser	Val	Thr	Ser	135 Ser	Ser	Leu	Cys	Val	140 Gly
Val	Asp	Ala	Asn	145 Arg	Asn	Trp	Asp	Ala	150 Gly	Phe	Gly	Lys	Ala	155 Gly	Ala	Ser	Ser	Ser	160 Pro
Cys	Ser	Glu	Thr	165 Tyr	His	Gly	Lys	Tyr	170 Ala	Asn	Ser	Glu	Val	175 Ile	Val	Lys	Ser	Ile	180 Val
Asp	Phe	Val	Lys	185 Asn	His	Gly	Asp	Lys	190 Lys	Ala	Phe	Leu	Ser	195 Ile	His	Ser	Tyr	Ser	200 Gln
Leu	Leu	Leu	Tyr	205 Pro	Tyr	Tyr	Tyr	Thr	210 Thr	Gln	Ser	Ile	Pro	215 Asp	Lys	Thr	Glu	Leu	220 Asn
Gln	Val	Ala	Lys	225 Ser	Ala	Val	Ala	Ala	230 Leu	Lys	Ser	Leu	Tyr	235 Gly	Thr	Ser	Tyr	Lys	240 Tyr
Gly	Ser	Ser	Ile	245 Thr	Thr	Ile	Tyr	Gln	250 Ala	Ser	Gly	Gly	Ser	255 Ile	Asp	Trp	Ser	Tyr	260 Asn
Gln	Gly	Gly	Lys	265 Tyr	Ser	Phe	Thr	Phe	270 Glu	Leu	Arg	Asp	Thr	275 Gly	Arg	Tyr	Gly	Phe	280 Leu
Leu	Ala	Ala	Ser	285 Gln	Ile	Ile	Pro	Thr	290 Ala	Gln	Glu	Thr	Trp	295 Leu	Gly	Val	Leu	Thr	300 Ile
Met	His	Glu	Thr	305 Val	Asn	Asn													

TABLE 9-7 Substrates for Carboxypeptidase A

Type I substrate: polypeptide

Type II substrate: N-acyl-tripeptide

Type III substrate: N-acyl-dipeptide or ester analog

Type IV substrate: unsubstituted dipeptide

Type V substrate: N-acyl-α-amino acid or O-acyl-α-hydroxy acid

simple acetyl group (acetyl-L-phenylalanine and acetyl-L-mandelate are reasonably good CPA substrates) to a polypeptide. However, there are some further points that should be noted. In all but Type V substrates, the penultimate residue is an α-amino acid which should have the L-configuration. The importance of this configurational preference increases with the bulk of the side chain, and there is also a preference for large hydrophobic side chains at this position.* Finally, the conversion of a Type IV substrate into a Type III substrate by N-acylation boosts its CPA-catalyzed hydrolysis

* These points have been tested only with peptide substrates, although they may hold for esters as well.

rate by two to four order of magnitude. The possible significance of this finding will be discussed later.

Since they are relatively easy to prepare or purchase, and because their hydrolysis is more readily and conveniently studied, substrates of Type III–V have received far more attention in CPA mechanism studies than have higher peptides or proteins.* It will be recalled that small synthetic substrates such as these have also been predominant with chymotrypsin studies, where they have led to no special difficulties. However, recent studies with larger peptide substrates[65] and the X-ray diffraction studies from Lipscomb's laboratories[63] reveal that the CPA-active site is quite large. It is apparently capable of interacting not only with the C-terminal and penultimate amino acid residues on either side of the scissile bond, but also with up to three additional residues on down the peptide chain in a polypeptide or protein substrate. (Esters have not been examined.)

This situation gives rise to possible complexities when dipeptide substrates are employed, for they may bind both productively and unproductively, or even two or more at once, within the large CPA-active site. Indeed it has been apparent for some time that CPA kinetics are not infrequently subject to such complications as substrate inhibition, substrate activation, product activation, and product inhibition. These kinetic complexities are not yet fully understood, but they may in large part be due to variations in the number and kind of binding modes various small substrates are susceptible to.[66]

Unfortunate though it may be, almost all of the kinetic and chemical evidence gathered in attempts to shed light on the mechanism of action of CPA involves experiments in which small dipeptide or ester substrates were employed, either directly or in the assays used to monitor the effects of chemical modification of the enzyme on its activity. Thus, although much of this evidence is quite useful, it must be interpreted with care and any conclusions drawn may be subject to correction or refinement as data are gathered on well-behaved larger substrates such as Type II, which seem to be less subject to complexities in binding.[67]

Particularly troublesome in this regard are the apparent differences observed in the behavior of CPA toward certain commonly employed N-acyl dipeptide substrates on the one hand and an ester analog on the other. (All of these test substrates exhibit complex kinetics indicative of multiple binding problems.) On the basis of these differences, it has been implied that CPA behaves differently toward peptides *in general* and esters *in general*.[66,68] Although this *may* be so, this idea has met with some skepticism[63,69] and must be regarded as highly tentative until comparisons are made using more

---

* The relatively few esters that have been reported as CPA substrates are mostly of Type V. No Type I or II esters appear to have been investigated.

different substrates (particularly more esters), preferably those that are well behaved kinetically.

Of particular interest with CPA is the fact that it is a true metalloenzyme. Native CPA contains one Zn atom at the active site, which is essential to enzymatic activity.[70] However, native zinc may be replaced in the enzyme by a number of other metals with varying effects upon the apparent activity of enzyme (Table 9-8). The X-ray diffraction studies of Lipscomb and co-workers[63] show that the zinc atom is bound by three enzymic ligands, the

TABLE 9-8 Relative Apparent Activities of Some Metallocarboxypeptidases[a]

Metal	Peptidase Activity[b]	Esterase Activity[c]
apo (no metal)	0	0
Zn	1.0	1.0
Co	1.6	0.95
Ni	1.1	0.87
Mn	0.08	0.35
Cd	0	1.50
Hg	0	1.16
Pb	0	0.52
Ca	0	0

[a] **From data of Coleman and Vallee.[70]**
[b] **Based on rate assays using the dipeptide substrate, carbobenzoxy-glycyl-L-phenylalanine (CGP). Carried out at 0°C at pH 7.5 in a 20mM EDTA buffer containing 0.1M NaCl. Initial substrate concentration 20mM.**
[c] **Based on rate assays using the ester substrate, hippuryl-DL-β-phenyl-lactate (HPLA). Carried out at 25°C at pH 7.5 in a 5mM Tris buffer containing 0.1M NaCl. Initial substrate concentration 10mM.**

side chains of His 69, Glu 72, and His 196. In the free enzyme, a water molecule appears to serve as a fourth ligand; but when glycyl-L-tyrosine (a poor substrate since it is not N-acylated) is bound at the active site in CPA crystals, this water ligand is displaced by the carbonyl oxygen atom of the glycyl residue. We will consider the possible mechanistic significance of this Zn-substrate interaction later.

The crystallographic work of the Lipscomb group combined with chemical modification studies from several laboratories[68,71–79] have identified several amino acid side chains that are present in or near the CPA-active site and which appear to play an important role in its catalytic action. In addition to the zinc atom and its accompanying ligands, which we have already mentioned, the most positive evidence concerns Tyr 248, Arg 145, and Glu 270.

Even before the X-ray work, a functional role for one or two of the 19 tyrosine residues in CPA was strongly suspected on the basis of chemical modification studies. Thus acetylation with acetylimidazole or acetic anhydride, nitration with tetranitromethane, iodination, and azo-coupling with 5-diazo-1H-tetrazole or *p*-azobenzenearsonate all result in CPA derivatives which behave quite differently from the native enzyme in the usual rate assays with GCP or HPLA as shown in Table 9-9. All of these reagents (with the possible exception of the diazotetrazole at high concentration) produce their effects by reacting with tyrosine residues in the enzyme. Where other types of residues are also modified, it has been demonstrated through the use of a competitive inhibitor such as β-phenylpropionate (which binds at the active site and protects the enzyme against the effects of these reagents) that tyrosine modification alone is functionally significant.

The later entries of Table 9-9 concern chemical modification studies on CPA directed toward arginine and glutamic acid side chains after the crystallographers had suggested possible functional roles for Arg 145 and Glu 270. Particularly worthy of special note among these more recent chemical modification studies is the final entry in Table 9-9. N-Bromoacetyl-N-methyl-L-phenylalanine **XI** is an affinity label for CPA, which behaves in much the

$$C_6H_5CH_2\!\!-\!\!\underset{\underset{\underset{\displaystyle CH_3}{|}}{\underset{\displaystyle N\!\!-\!\!\overset{\displaystyle O}{\overset{\|}{C}}\!\!-\!\!CH_2Br}{\diagdown}}}{\overset{\displaystyle H}{\overset{\diagup}{C}}}\!\!\rightharpoonup\!\! CO_2H$$

XI

same way toward this enzyme that TPCK behaves toward chymotrypsin. It irreversibly inactivates CPA toward both peptides and esters by specifically alkylating the carboxyl group of Glu 270 in the enzyme-active site. The inactivation rate is retarded by competitive inhibitors, as expected for an active-site-directed reagent. Also, as noted previously with TPCK and chymotrypsin, the kinetics of inactivation indicate that a noncovalent CPA-XI complex is formed in a reversible binding step prior to the actual alkylation.

The most definitive single source of information concerning the identity and possible function of amino acid side chains at the active site of CPA is the X-ray diffraction data on a crystalline complex between the enzyme and the substrate glycyl-L-tyrosine.[63] Translated into structural terms, these data elucidate four sites of enzyme-substrate interaction. First, the tyrosyl side chain of the substrate occupies a hydrophobic pocket in the active site

TABLE 9-9 Effects of Chemical Modification on the Apparent Peptidase and Esterase Activities of Carboxypeptidase A

Reagent (Ref.)	Residue(s)[a] Modified	Protection[b]	Peptidase Activity[c] (CPA_{mod}/CPA_{native})	Esterase Activity[c] (CPA_{mod}/CPA_{native})	Residue Identities[d]
Acetyl imidazole[71]	2 Tyr	+	0.02	7.0	Tyr 198 Tyr 248
Acetic anhydride[72]	2 Tyr	+	0.02	6.1	Tyr 198 Tyr 198
Iodination[73] (I_3^- or ICl)	5–6 Tyr or more	+	0.02	5 to 6	Tyr 248
Tetranitromethane[74]	1.2	+	0.1	1.7	Tyr 198 Tyr 248
5-Diazo-1H-tetrazole (8X)[75]	1.1 Tyr	+	0.9	1.8	Tyr 198
5-Diazo-1H-tetrazole (30X)[75]	2 Tyr + 1 His	+	0.05	1.8	Tyr 198 Tyr 248 His ?
2-3-Butanedione (diacetyl)[76]	1 to 2 Arg	−	0.1	2.4	Arg 145 (Arg 71, Arg 127)
Cyclohexyl-3-(2-morpholino-ethyl)carbodiim-ide metho-p-toluene sulfonate[77]	?	+	0	0	Glu 270
N-Ethyl-5-phenyl-isoxazolium-3-sulfonate[78]	1 Glu	+	0	0	Glu 270
N-Bromoacetyl-N-methyl-L-phenylalanine[79]	1 Glu	+	0	0	Glu 270

[a] In cases where one or more of the residues modified were demonstrated to have no functional importance, these residues have been omitted from the table.

[b] Indicates whether or not protection against activity changes is afforded by competitive inhibitors such as β-phenylpropionate.

[c] Conditions essentially as in Table 9.8.

[d] Positive identities from modified peptide isolation and characterization are underlined. Tentative identifications based on X-ray data alone are not (see Ref. 63).

which appears large enough to accommodate an even larger group (e.g., a tryptophan side chain). Second, the free terminal α-carboxylate group of the substrate forms an ion pair with the guanidinium side chain of Arg 145. Third, as mentioned earlier, the carbonyl group of the peptide bond becomes the fourth zinc ligand in place of water. Fourth, the free terminal α-amino group of the substrate glycyl residue interacts (through an intervening water molecule) with the carboxylate group of Glu 270. This final interaction would not be possible for a good substrate in which the α-amino group of the penultimate residue would be acylated. In fact, it is suggested that this substrate-Glu 270 interaction is specifically responsible for the stability of the Gly-Tyr-CPA complex against hydrolysis, allowing crystallographic measurements to be made. Implicit in this suggestion is a possible role for Glu 270 as a nucleophile or proton acceptor in the catalytic action of CPA toward good substrates.

When the structure of the Gly-Tyr-CPA complex is compared with that of the native enzyme,[63] it is apparent that the binding of Gly-Tyr brings about some rather extensive conformational changes within the protein in the region of the active site. These include (1) a disengagement of the Arg 145 guanidinium group from its hydrogen bond with the peptide chain carbonyl at residue 155 and a movement of some 2Å to its position in the enzyme-substrate ion pair, (2) a 2 Å repositioning of Glu 270, and (3) a disengagement of Tyr 248 from its interaction with Glu 249 in the free enzyme so that its OH group moves some 12 Å through an arc of about 120°, coming to rest just 2.7 Å from the peptide nitrogen of the bound substrate. Furthermore, substrate binding transforms the whole active site cavity of CPA from a water-filled region, freely accessible to the solvent, to a hydrophobic region, closed off to the solvent. Most of the several water molecules which before binding had participated in solvation of both the substrate molecule and active site amino acid side chains are expelled when binding occurs. This whole process provides the clearest and most dramatic example known of the kind of enzyme-substrate interaction Koshland had in mind when he proposed the induced-fit theory,[80] in which enzyme active sites were in general envisaged as somewhat flexible and undefined before binding occurred, locking active site residues into defined positions around the substrate.

As noted earlier, kinetic studies on CPA using small synthetic peptide and ester substrates are often complicated by effects that are not well understood, but which are probably due in large part to multiple and/or unproductive substrate binding modes. Fortunately, however, a few substrates have been investigated which give rise to straightforward Michaelis-Menten kinetics with CPA. Before we consider mechanistic arguments, some of the findings from kinetic studies on some of these presumably well-behaved substrates merit our attention.

The simplest behavior is observed with a group of N-acylated tripeptide substrates such as carbobenzoxy- or benzoylglycylglycyl-L-phenylalanine, -leucine, or -valine recently studied by Auld and Vallee.[67,81] The pH dependence of the simple Michaelis-Menten parameters* (k_{cat}, K_m, and k_{cat}/K_m) clearly indicate that substrate binding for these tripeptides depends on a single ionization of $pK_a \sim 9.0$, whereas catalysis per se (i.e., k_{cat}) depends on a single ionization of $pK_a \sim 6.2$. Acylation of tyrosyl residues in CPA using acetylimidazole results in very little effect on K_m for these substrates, but it does reduce k_{cat} by an order of magnitude. Moreover, the rates of CPA-catalyzed hydrolysis of BzGlyGly-L-Phe (under conditions where [S] $\gg K_m$) in D_2O and in H_2O are essentially identical within the limits of experimental error. This indicates the absence of a D_2O kinetic solvent isotope effect on k_{cat}.

Somewhat different behavior is observed with two ester substrates, O-acetyl-L-mandelate[69] and O-(*trans*-cinnamoyl)-L-β-phenyllactate,[82] which give rise to Michaelis-Menten kinetics complicated only by competitive product inhibition (by L-mandelate and L-β-phenyllactate, respectively). The pH dependence of the Michaelis-Menten parameters for these esters has been determined, but in neither case is a straightforward interpretation in terms of independent effects of particular ionizations upon binding and catalysis possible. However, k_{cat}/K_m *versus* pH profiles for both esters are bell shaped, implying two functionally essential ionizations whose pK_a values in the free enzyme are 6.9 and 7.5 for O-acetyl-L-mandelate and 6.5 and 9.4 for O-(*trans*-cinnamoyl)-L-β-phenyllactate. The latter pair of pK_a values is quite similar to those observed with the tripeptide substrates and it is possible that the same ionizations are involved in both cases. However, the second pK_a of 7.5 for O-acetyl-L-mandelate appears to be anomalous, and it may represent an ionization which does not affect the other substrates. Another difference between these two ester substrates is found in their behavior with acetylated CPA. Tyrosyl acetylation abolishes the activity of CPA toward O-acetyl-L-mandelate.[83] On the other hand, acetylation has no effect on K_m for O-(*trans*-cinnamoyl)-L-β-phenyllactate while reducing k_{cat} by a factor of about 2.[84] Here again, the behavior of the latter ester is not unlike that of the tripeptides (although the magnitude of the effect of acetylation on k_{cat} is somewhat less), whereas the behavior of the former ester is anomalous. It has been suggested[85] that L-mandelate esters, which are in fact much poorer substrates for CPA than L-β-phenyllactate esters, be considered nonspecific substrates for CPA.

Finally, in contrast with the findings for tripeptide substrates, a significant D_2O kinetic solvent isotope effect has been observed in the CPA-catalyzed hydrolysis of the "specific" ester substrate O-(*trans*-cinnamoyl)-L-β-phenyllactate.[86] Although a change in solvent from H_2O to D_2O does not affect K_m for this substrate or K_i for its inhibitory product, the $k_{cat}^{H_2O}/k_{cat}^{D_2O}$ ratio is approximately 2.

---

* See Chapter 5 for a discussion of pH effects on enzyme kinetics.

b. Mechanism of Action

We are now in a position to construct some meaningful mechanistic hypotheses concerning the previously discussed (1) specificity requirements and strong preferences of CPA, (2) metal-ion requirement (Table 9-8), (3) chemical modification experiments (Table 9-9), (4) X-ray diffraction studies on a crystalline complex between CPA and the substrate Gly-L-Tyr, and (5) kinetic studies on well-behaved substrates. To avoid undue confusion, we will proceed on the assumption that productive interaction between CPA and its specific substrates, be they peptides or esters, may be characterized in terms of a single binding mode.*

It is reasonable to suppose, as we did with chymotrypsin, that the free energy of association between a good CPA substrate and the enzyme in a productive complex is largely attributable to hydrophobic bonding between the side chain of the C-terminal residue of the substrate and the hydrophobic pocket of the CPA-active site. This interaction is clearly observable in the structure of the crystalline complex of CPA with Gly-L-Tyr and it would account for the strong preference of CPA for substrates with a large hydrophobic side chain on the C-terminal residue.

When the substrate is bound at the active site, the obligatory terminal carboxylate ion of the substrate exerts a coulombic attraction toward the guanidinium ion of the side chain of Arg 145. The movement of this arginine residue toward the carboxylate group of the substrate to form an ion pair triggers the extensive conformational changes within the CPA-active site, which we enumerated earlier. Once these conformational changes are complete, a substrate with the L-configuration in the C-terminal residue will be positioned so that the carbonyl group of the scissile peptide or ester bond becomes a zinc ligand and the unsubstituted peptide nitrogen or ester oxygen can interact through hydrogen bonding with the hydroxyl group of Tyr 248. All of these interactions are indicated schematically in Figure 9-5. Furthermore, the carboxylate group of the Glu 270 side chain, which chemical modification studies have shown to be essential to catalysis, is in close proximity to the carbonyl group of the scissile peptide or ester bond. Moreover, Hartsuck and Lipscomb[63] have suggested that an additional constraint on substrate positioning may arise from a second hydrogen-bonding interaction between

---

* It must be recognized that this is probably an oversimplification. The many apparent discrepancies among various substrates in kinetic behavior and response to chemical modifications and metal ion substitutions have led some investigators to propose a "dual-site" model for CPA action.[87] This model is conceptually attractive since it can accommodate, albeit only schematically at the present time, many of the anomalies that have proved troublesome in the formation of mechanistic arguments concerning the action of CPA. On the other hand, meaningful assessments of the roles of individual enzymic groups in catalysis or of the nature of bond-making and bond-breaking events are not yet feasible in terms of this dual-site model, and we will not consider it here.

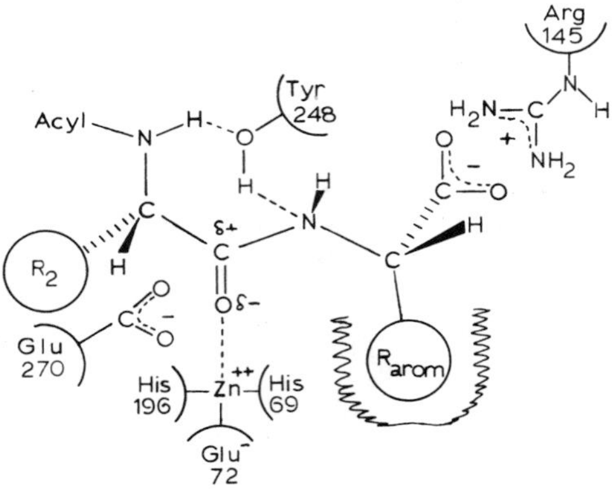

FIGURE 9-5. Schematic drawing indicating possible interactions between the C-terminal dipeptide of a CPA substrate and side chain groups present at the CPA active site.

Tyr 248 and the substrate; this involves the acylated α-amino group of the penultimate residue as indicated in Figure 9-5.

It is not yet clear what pathway the actual hydrolysis of the peptide bond takes once the substrate is bound, but some feasible possibilities have been advanced. First, it is certainly true that any pathway will require a nucleophilic attack on the carbonyl group of the scissile bond, either by water directly or by some nucleophilic group of the enzyme active site. Whatever the nucleophile, it is to be expected that the interaction between the positively charged zinc ion and its substrate ligand, the oxygen atom of the peptide or ester carbonyl group, should markedly enhance the polarization of that carbonyl group as indicated in Figure 9-5. This would in turn enhance the reactivity of the carbonyl carbon toward an attacking nucleophile, particularly since the reaction will be taking place in a generally hydrophobic environment (provided by the substrate-occupied CPA active site as discussed earlier) where electrostatic effects are intensified. Thus it is likely that the zinc ion of CPA has a dual role in catalysis, both helping to position the peptide carbonyl group for efficient nucleophilic attack and helping to enhance its reactivity.

A central role in catalysis for the carboxylate side chain of Glu 270 is strongly implied by both chemical modification studies and the X-ray diffraction studies. One possibility is that the function of this carboxylate group is much the same as the function we have postulated for the imidazolyl group of His 57 in chymotrypsin deacylation. This possibility is illustrated schematically in Eq. 9-13, which shows general-base catalyzed attack by water on the activated carbonyl of the substrate (where X is the alcohol or

$$(9\text{-}13)$$

amine leaving group) followed by general acid-catalyzed breakdown of the tetrahedral intermediate. A second possibility, closely related to the first, is that general base catalysis by Glu 270 is accompanied by general acid catalysis by another active-site functional group, perhaps Tyr 248. Equation 9-14 represents this possibility as a concerted process, although it might occur

$$(9\text{-}14)$$

stepwise through a tetrahedral intermediate much as in Eq. 9-13. A third possibility is that Glu 270 functions as a nucleophile rather than as a general base, attacking the activated carbonyl of the substrate to form an anhydride intermediate, which is subsequently hydrolyzed by water. This is shown in Eq. 9-15. A role for a general acid is again indicated in this mechanism, but the proton donor might be either an enzyme group or a water molecule.

Let us now recount the major points of evidence that bear upon the mechanism of CPA action and see to what extent our mechanistic hypotheses accommodate these data.

We have accounted for most of the specificity requirements of CPA in our description of the initial binding interaction between CPA and its substrates, as exemplified by the crystallographic data on the CPA-Gly-L-Tyr complex. In particular we noted three major interactions which are associated with three major features of specificity: (1) hydrophobic bonding involving

$$(9\text{-}15)$$

the side chain of the C-terminal residue of the substrate; (2) ion pairing between $Arg^{\oplus}$ 145 and the obligatory C-terminal carboxylate ion of the substrate; and (3) coordination of the scissile carbonyl group with the enzyme metal ion. All these interactions are possible only when the C-terminal residue of the substrate has the L-configuration.

Concerning the metal-ion requirement of CPA, we have postulated an essential role for zinc in both binding and catalysis.

Regarding the results of chemical modification experiments and X-ray diffraction studies on the identification of probable functionally important residues at the CPA active site, we have assigned essential roles to Glu 270 in catalysis and to $Arg^{\oplus}$ 145 in binding. We have also mentioned the possibility that Tyr 248 may participate either in binding through hydrogen bonding or in catalysis as a general acid, but neither of these roles can be regarded as essential on the basis of current evidence (see below).

Finally, we must consider the evidence derived from kinetic studies with well-behaved substrates. From the pH dependence of the kinetics for tripeptide substrates and the ester, O-(*trans*-cinnamoyl)-L-β-phenyllactate, we infer the importance of ionizing groups with pK_a of values ~ 6.5 and ~ 9.0. It seems highly likely that the lower pK_a may be attributed to Glu 270, which has been shown to have a pK_a of 6.5–7.0 in affinity-labeling experiments.[79] The tripeptide kinetic studies indicate that this group participates only in catalysis, and this is the role we have portrayed for it. The higher pK_a is more difficult to account for in unequivocal terms. It could possibly be attributed to Tyr

248.* However, the tripeptide kinetic studies indicated that for those substrates at least, the $pK_a \sim 9$ ionization affected only substrate binding, and yet tyrosyl acetylation had essentially no effect on binding, either for these tripeptide substrates or for O-(*trans*-cinnamoyl)-L-β-phenyllactate. The most likely alternative possibility is the pK_a of the water ligand of the active site zinc ion.[84,85] This water ligand is displaced from the metal ion upon substrate binding, and such displacement would be much more difficult if ionization to the hydroxide complex had occurred. If this interpretation is correct, however, it is somewhat surprising that the observed pK_a is apparently insensitive to metal ion substitutions.[81,84] On the other hand, there are apparently no adequate models among known metal ion complexes with which to compare the behavior of CPA in this regard.

Finally, concerning the mechanistic possibilities implied by Eq. 9-13 to 9-15, we note that acyl group reactions which involve proton transfer between oxygen or nitrogen and a general acid or base catalyst, along with bond making or breaking between heavier atoms, commonly proceed some two to three times faster in water than in D_2O.[88] Thus, taken at face value, the absence of a D_2O kinetic solvent isotope effect on k_{cat} for the tripeptide substrates would seem to eliminate Eq. 9-13 and 9-14 from serious consideration since general-acid and/or base catalysis is involved in every step. Furthermore, the anhydride pathway of Eq. 9-15 can accommodate both the absence of an isotope effect for peptides (assuming that the initial nucleophilic attack by the carboxylate ion of Glu 270 is rate limiting) and an isotope effect of ~ 2 for O-(*trans*-cinnamoyl)-L-β-phenyllactate (assuming rate-limiting hydrolysis of the anhydride intermediate).[85] The anhydride pathway is also attractive from the point of view of steric considerations based on the X-ray work[63] and on the basis of certain model system studies.[85] Unfortunately, however, though the data at hand do favor Eq. 9-15, it is not possible to rule out mechanisms such as those implied by Eqs. 9-13 and 9-14 on the basis of the kinetic isotope effect findings. Reactions involving proton transfers in rate-limiting steps and yet giving rise to essentially no kinetic isotope effects are known.[88] Furthermore, solvent isotope effects in enzyme-catalyzed reactions are especially difficult to interpret for a number of reasons[88] and must be regarded with caution.

It should be apparent that the anhydride pathway of Eq. 9-15 is related to the acyl-enzyme mechanism for chymotrypsin and other serine proteases

---

* It has been suggested that the "abnormal" pK_a of ~ 7.5 observed with the nonspecific ester substrate, O-acetyl-L-mandelate may be attributed to Tyr 248,[85] since this is the only known well-behaved substrate toward which acetylation *completely* abolishes activity. An abnormally low pK_a for an "essential" tyrosyl residue in CPA has also been suggested on the basis of the pH dependence of CPA modification using tetranitromethane.[74]

in that it implies a covalent enzyme-substrate intermediate in the catalytic pathway. In principle then, using the same kinds of experimental approaches that have elucidated the acyl-enzyme pathway for the serine enzymes, one might expect to obtain evidence for the anhydride intermediate of CPA-hydrolysis. In fact, some of these approaches have been tried with CPA,[89-91] and no evidence for a covalent intermediate has been found. Of course failure to obtain positive evidence for a certain hypothesized mechanistic pathway cannot be interpreted to mean that the pathway does not in fact exist. In this particular case, an enzyme-substrate anhydride intermediate would be much less stable than the acyl-enzyme intermediate of the serine enzymes and thus one would expect it to be a relatively elusive species.[85] We shall have to await further developments in the ongoing investigations of CPA action for clarification of the many points of uncertainty surrounding current mechanistic hypothesis. Nonetheless, the progress that has been made toward understanding the mechanisms of action of this enzyme is impressive.

Lysozyme (EC 3.2.1.17)

In 1922, seven years before he was to make the discovery of the antibiotic penicillin which made him famous, Alexander Fleming found that his own nasal mucus contained an agent capable of killing certain bacteria by dissolving, or lysing, their cell wall structure.[92] This agent, which has since been found widely distributed in the tissues and secretions of a number of animals, both vertebrate and invertebrate, as well as in plants, bacteria, and phages,[93] turned out to be the enzyme lysozyme. The whites of eggs, which contain particularly large amounts of this enzyme, have received the most study, begun by Fleming himself some 50 years ago.[92]

Hen egg white lysozyme is relatively easy to isolate and purify. It was first crystallized in 1937[94] (the same year as carboxypeptidase A), which puts it among the earliest enzymes available for study as pure proteins. Indeed, it has been studied extensively since those early days of enzyme and protein investigation. But until relatively recently most of that study focused primarily on its properties as a *protein* rather than as an enzyme. Its molecular weight (14,500), electrophoretic properties (pI $\sim$ 10.5), and crystallographic properties were all known before 1950.[95] In 1963, two groups working independently announced the complete, 129-amino acid primary structure of lysozyme.[96,97] And by 1965, lysozyme became the second protein in history (the first was myoglobin in 1962) to have its complete and detailed molecular structure elucidated by X-ray crystallography.[98] On the other hand, before 1957, when Berger and Weiser found that the enzyme degrades chitin as well as bacterial cell walls,[99] it was not even known what sort of chemical linkage lysozyme attacks. And it was not until 1966 that the actual structure of the glycopeptide polymer which is the substrate of lysozyme in bacterial cell walls was fully elucidated.[100] Nevertheless, primarily due to the enormous contribution to our understanding provided by the X-ray crystallographic work, and to recent work of a more classical sort stimulated by the conjectural conclusions of the crystallographers, we now are as close to a viable "mechanism for action" for lysozyme as we are for chymotrypsin, carboxypeptidase, or any of a number of other enzymes which have received many times more enzymological study than lysozyme.

Figure 9-6 shows the structures of chitin and the bacterial cell wall glycopeptide which are the known substrates for lysozyme. Lysozyme catalyzes

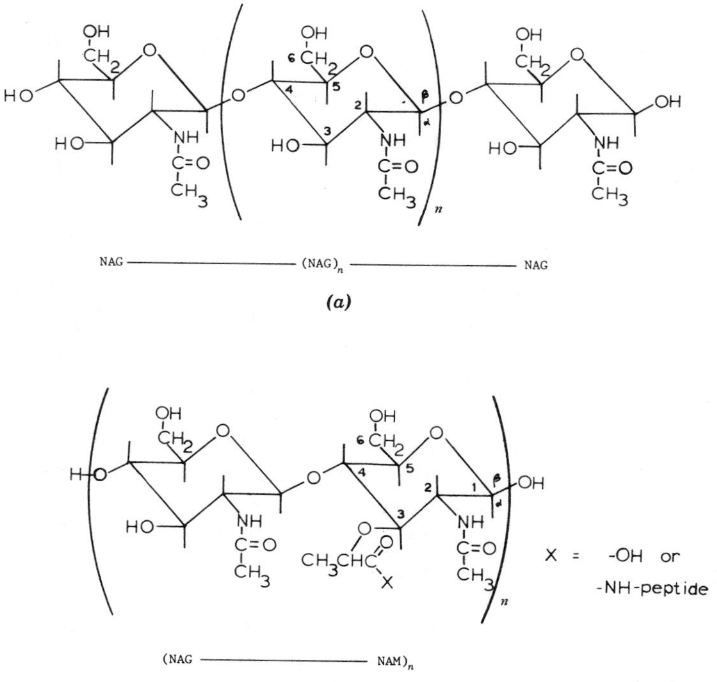

FIGURE 9-6.   (a) The structure of chitin or poly-NAG, which is a linear chain of repeating N-acetyl-D-glucosamine (NAG) units linked together by β(1 → 4) glycosidic linkages. (b) The repeating unit of the cell-wall glycopeptide polymer, poly-alternating-NAG-NAM. This polymer differs from poly-NAG in that every other glucosamine unit has a D-lactic acid residue attached to it by an ether linkage at C_3, giving rise to the saccharide units known as N-acetylmuramic acid (NAM). At random intervals along the chain, short polypeptide crosslinks extend from the lactate group of an NAM residue to its counterpart in another chain, thus forming the interwoven network of the cell-wall material.

the hydrolytic cleavage of $\beta(1 \rightarrow 4)$ glycosidic linkages in these polymeric substrates so long as the C_4 atom belongs to an NAG (N-acetyl-D-glucosamine) residue. Thus cleavage can occur between any two residues in chitin or oligosaccharides derived therefrom, but in cell wall polymers or oligosaccharides it can occur only between an NAM (N-acetylmuramic acid) C_1 and the C_4 of the adjoining NAG residue. In all cases, cleavage occurs between C_1 and the ether oxygen; never at C_4. Poly-NAM is not a lysozyme substrate for reasons we will discuss shortly.

The X-ray crystallographic studies on lysozyme by Phillips, North, Blake, and their collaborators[98,101–105] led of course to a complete structural characterization of this enzyme, the details of which we will not concern ourselves with here. An excellent, superbly illustrated presentation of the structure of lysozyme and the other enzymes discussed in this chapter may be found in a recent monograph by Dickerson and Geis.[106] Suffice it to say that the molecule itself assumes roughly a prolate spheroidal shape, 45 Å from top to bottom and 30 Å through at the middle, with a deep cleft running approximately halfway around it at the equator. The nonpolar hydrophobic amino acid side chains are concentrated toward the interior of the molecule and along the bottom of the cleft, while most of the polar residues are distributed about the enzyme's surface.*

Of primary interest to us are the insights concerning the possible mechanism of action of lysozyme which the crystallographers gained from studies of crystalline complexes of the enzyme with saccharide inhibitors bound at the active site, and from model building in which various hypothetical enzyme-substrate interactions could be tested on simple steric grounds. From such considerations, the crystallographers deduced that the entire cleft extending halfway around the molecule as described above constituted the active site of lysozyme. This cleft is just long enough to interact simultaneously with six sugar residues when a poly-NAG or poly-alternating-NAG-NAM chain is wrapped around the enzyme molecule as indicated in Figure 9.7. A number of hydrogen-bond and van der Waals contact interactions between these six sugar residues and appropriate backbone or side chain atoms of the enzyme can be identified or postulated. Some of these interactions are indicated in Figure 9-8. With the binding of substrates or inhibitors, there is a small movement of the "top" side of the cleft toward the substrate, slightly narrowing the cleft. For example, Trp 62 moves some 0.75 Å (see Figure 9-8). This may be another example of Koshland's induced-fit[80] phenomenon, although it is by no means as impressive as the example provided by carboxypeptidase A.

---

* This kind of disposition of hydrophobic and polar residues seems to be common to all proteins whose structures have been determined by X-ray diffraction, and probably will turn out to be a common feature of globular proteins in general.[107]

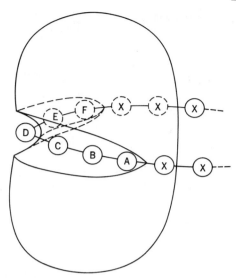

FIGURE 9-7. A highly stylized representation of the lysozyme molecule with a polysaccharide substrate, (NAG)$_n$ or (NAG-NAM)$_n$, occupying the active-site cleft of the enzyme. The cleft accommodates just six sugar residues (represented here as beads on a string) in six specific subsites, A through F. The NAM residues of cell-wall polysaccharides will fit only in subsites B, D, and F. The cleavage point lies between subsites D and E. The residues designated X interact insignificantly, if at all, with the enzyme.

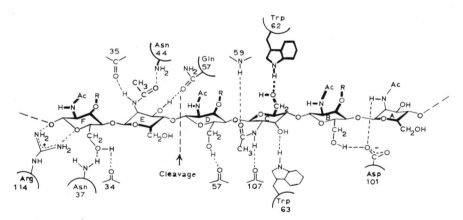

FIGURE 9-8. A schematic representation of the hydrogen-bond interactions of a bound lysozyme substrate with enzyme groups in the active-site cleft. The view is from outside the enzyme looking toward the bottom of the cleft, which extends from right to left. The dark, heavy edges of the pyranose rings are exposed while the light edges are buried against the bottom of the cleft. (After Dickerson and Geis,[106] p. 76).

A finding of special significance was the observation from model building studies that subsite D will not accommodate a pyranose ring in its normal chair conformation because of steric interactions of the protein backbone at that subsite with the C_6 hydroxymethyl group. It was suggested that this steric effect forces the D ring to distort toward a half-chair conformation when saccharide residues of a substrate oligosaccharide are bound in the active-site cleft on either side of subsite D. Another steric effect is encountered in subsites A, C, and E, where there is not enough room in the cleft to accommodate a substituted-OH group on C_3 of the pyranose ring. Thus NAM residues of cell wall polysaccharides, with their C_3 lactyl ether groups, are allowed access only to subsites B, D, and F where the lactyl side chains (and occasional peptide crosslinks) point out away from the cleft as in Figure 9-8.

Since $(NAG-NAM)_n$ substrates are cleaved only at the C_1 atom of NAM residues, and since NAM residues can occupy only subsites B, D, and F, it could be inferred that the cleavage point must be either between subsites B and C, or between D and E. However, the B–C possibility was eliminated on the grounds that the trisaccharide $(NAG)_3$, which is an inhibitor of lysozyme, forms a very stable nonproductive complex with the enzyme in which it specifically occupies subsites A, B, and C.* Having thus identified the cleavage point between residues D and E, the crystallographers were able to implicate the carboxyl side chains of Glu 35 and Asp 52 as the only two functionalities of potential catalytic significance close enough to the scissile $\beta(1 \rightarrow 4)$ linkage to be involved in the cleavage mechanism. These two residues take up positions on either side of the bond to be cleaved. The carboxyl group of Glu 35 is in a nonpolar, hydrophobic environment deep within the cleft beneath the scissile bond when a substrate is in place. It appears to be the one carboxyl group of lysozyme with an abnormally high pK_a, so that it would be largely protonated in the region of maximal lysozyme activity between pH 5 and 6. Hence it might be expected to serve as a proton donor toward the oxygen on the leaving-group side of the scissile bond from which its nearest oxygen atom is only 3 Å distant. The carboxyl group of Asp 52, on the other hand, is in a much less hydrophobic environment on the upper lip of the cleft above the scissile bond as viewed in Figure 9-7 or 9-8. It is presumably ionized in the operational pH range of lysozyme, and in fact Blake et al.[101] point out that it is in a position to accept hydrogen bonds from a pair of nearby asparagine residues, which may confer upon it an abnormally *low* pK_a. It has been suggested that this pK_a value for Asp 52 may be as low as 2.[109] However, a very recent investigation of the ionization

* X-ray diffraction studies on the crystalline lysozyme-$(NAG)_3$ complex were the primary source of information leading to most of the deductions and inferences cited in this discussion of lysozyme-substrate interactions.

behavior of the catalytic carboxyl groups of lysozyme assigns effective macroscopic ionization constants of 4.5 and 5.9 to Asp 52 and Glu 35, respectively, in $0.15M$ KCl at $25°$.[108] Model building based on the X-ray data suggests that the closest oxygen atom of Asp 52 is only about 3 Å distant from either the C_1 carbon atom or the adjacent ring oxygen atom of the D-ring of a bound substrate. Thus it could interact with either of these two substrate atoms during catalysis.

On the basis of the foregoing X-ray and model-building studies and what was known about the mechanisms of nonenzymic glycopyranoside hydrolysis, a mechanism for lysozyme action was proposed.[110] It was known that in hydroxylic solvents and in the absence of powerful nucleophiles, nonenzyme-catalyzed substitution at the C_1 carbon of a glycopyranoside tends to proceed by rate-determining cleavage of the exocyclic carbon-oxygen bond to give a cyclic carboxonium ion intermediate **XII**. This ion assumes a

XII

conformation with C_2, C_1, O, and C_5 coplanar, allowing maximum overlap of the p-orbitals of the sp^2-hybridized carbon and the oxygen. Substitution is completed by rapid attack of the nucleophile (e.g., water) on carbon atom C_1. The reaction is catalyzed by acids which function to protonate the alkoxide leaving group prior to bond cleavage.*

In the enzyme, it was assumed that all of the preceding features of the hydrolysis mechanism were preserved and that the enzymatic rate enhancement might be attributed to various factors suggested by the structural characterization of the interaction of the lysozyme active site with specific substrates. It was suggested that the carboxyl of Glu 35 functions as a *general* acid catalyst, assisting glycosidic bond rupture by protonating the leaving

---

* The question concerning the timing of this protonation with respect to bond cleavage was not dealt with in detail until recently. However, Bruice and co-workers[111-113] have determined that the hydrolysis of glycopyranosides with ordinary alcohol leaving groups proceeds with exclusive *specific acid catalysis*, implying that protonation must be complete before the rate-limiting carbon-oxygen bond cleavage. This finding holds even for model systems in which ground-state planarity for the C_2, C_1, O, C_5 bond system is provided, and there is no reason to believe it would not hold for enzyme-catalyzed glycosidic bond rupture as well.

group in the rate-limiting step.* The carboxylate ion of Asp 52 was pictured as stabilizing the developing positive charge of the carboxonium ion in the transition state for bond rupture. The possibility that it might serve as a nucleophile to displace the disaccharide in subsites E and F, giving a glycosyl-enzyme intermediate which would subsequently be attacked by water, was rejected on the grounds that model building showed that the requisite geometry for such a direct S_N2-type displacement would be impossible. The leaving group would have to depart on a path which is blocked by the structure of the enzyme. Moreover, constraints on Asp 52 do not allow it to approach carbon atom C_1 in subsite D more closely than about 3 Å, which is somewhat greater than a normal carbon-oxygen bond distance.

In addition to the acid catalysis provided by Glu 35 and the electrostatic catalysis provided by Asp 52, it was proposed that the distortion away from the stable chair conformation of the pyranose ring upon binding in subsite D was a possible example of catalysis by induction of strain (see Chapter 7). If a half-chair conformation as in **XII** is assumed by the carboxonium ion intermediate, the transition state leading to this intermediate must also be distorted from the normal chair conformation for six-numbered carbocycles. To the extent that this distorted conformation is forced upon the pyranose ring in subsite D by the enzyme in the ground-state–enzyme-substrate complex, the activation energy barrier between that ground state and the transition state might be reduced.

In summary, then, the sequence of events describing the lysozyme-catalyzed hydrolysis of poly-(NAG-NAM) was proposed to be:

1. The substrate is bound in subsites A to F by hydrogen bonds and other forces, and in the process the NAM residue in subsite D is distorted toward the conformation it must assume in the transition state for glycosidic bond cleavage.

2. A proton is transferred from Glu 35 to the oxygen connecting C_1 of the NAM residue in subsite D and C_4 of the NAG residue in subsite E.

3. Rupture of the scissile bond occurs with the developing positive charge of the carboxonium ion intermediate being stabilized by the Asp 52 carboxylate ion.

* The possible role of Glu 35 as a proton donor toward the leaving group remains highly plausible. But recent evidence briefly alluded to in an earlier footnote and discussed in greater detail later in the text makes it seem unlikely that general acid catalysis is involved; rather, specific acid catalysis with proton transfer complete before bond rupture is involved. Nonetheless, under conditions at the lysozyme-active site, Glu 35 might indeed be the *only* available proton donor. In any case, a rate enhancement attributable to stabilization of the cationic transition state by the carboxylate of Glu 35 even *after* the proton is transferred is to be expected, especially in the hydrophobic environment of lysozyme active-site cleft.

4. The fragment occupying subsites E and F diffuses away and is replaced by a water molecule or other nucleophilic acceptor while the fragment in subsites A to D remains associated with the enzyme. The acceptor nucleophile then attacks C_1 of the carboxonium ion in subsite D to complete the reaction. The original β-configuration about that C_1 carbon atom is retained in the product.

The structural characterization of lysozyme by X-ray crystallography and the deductions concerning the mechanism of action of that enzyme were presented essentially as recounted above in 1966.[101,104,110] It is worth noting again that at that time, the amount of corroborative experimental evidence concerning the behavior of lysozyme as an enzyme in solution was relatively meager compared with other enzymes such as chymotrypsin and carboxy-peptidase whose structures were being worked on at the same time. But following the X-ray work on lysozyme, and largely stimulated and guided by that work, various investigators carried out appropriate solution studies with lysozyme. The fact that the conclusions drawn from later work in large part are highly consistent with the initial conjectures of the X-ray crystallog-raphers concerning the nature of lysozyme catalysis provides dramatic support for the power and validity of X-ray crystallography as a technique for the study of enzyme action. Let us consider briefly some of the major findings supporting the crystallographers' hypotheses.

Determinations of binding constants for a large number of saccharide substrates and inhibitors with lysozyme in solution have been carried out, and collectively these data are impressively consistent with the six-subsite model.[114] By using the binding constant data, it has been possible to estimate the free energy of binding at each of the six subsites, and the estimated values are all negative (favorable to binding) with the important exception of that for subsite D, which is positive by an estimated 3–6 kcal/mole.[114] This is of course consistent with the crystallographers' inference concerning steric hindrance and conformational distortion of the pyranose ring occupying subsite D. It also allows an estimate of the magnitude of catalytic rate en-hancement that might be realized as a result of such a distortion. If the entire 3–6 kcal/mole were expended toward distorting the substrate partway toward the transition state, it might accelerate bond cleavage in the enzyme-substrate complex by a factor of 10^2 to 10^4.

Another point in the crystallographic model that has been confirmed by solution studies is the position of bond cleavage and its stereochemistry. Hexasaccharides like $(NAG)_6$ are cleaved nearly quantitatively between the second and third residues from the reducing terminus, which is as expected for cleavage between subsite D and E of the model if these substrates occupy all six subsites.[114] Further, the finding that lysozyme catalyzes not only hydrolysis (transfer of an oligosaccharide fragment to water) but also trans-glycosylation (transfer of an oligosaccharide fragment to acceptor saccharides or to other nonwater nucleophiles) allowed investigators to study the stereochemistry of bond formation by the acceptor molecule* and the specificity of subsites E and F toward acceptor structure.[114] The very fact that trans-glycosylation occurs implies that the reaction proceeds in a stepwise manner, and that a "glycosyl-enzyme" intermediate of some sort exists long enough for the leaving group to diffuse away and be replaced by an appropriate acceptor molecule before water can react with the intermediate. As predicted by the model mechanism described earlier, the acceptor must attack the glycosyl-enzyme intermediate along the same path taken by the departing cleavage product, and stereochemistry about the C_1 carbon atom is retained. That is, the final product is always formed with a $\beta(1 \rightarrow 4)$ glycosidic linkage.[115,116]

Very recently, the kinetic ramifications of the X-ray model have been worked out by Chipman[117] and have been found to account quite satisfactorily for the rather complex kinetics of lysozyme catalysis. Among the more interesting conclusions from these kinetic studies are the following: (1) substrates with fewer than six sugar residues are poorer substrates than those with six or more residues by a factor of some 10^4 to 10^6, because the former bind principally in an unproductive mode; and (2) the ratio of k_T/k_H (the specific rate constant for attack of a $(NAG-NAM)_n$ acceptor as opposed to attack by water on the glycosyl enzyme intermediate) is about 3×10^3. The second finding means that subsites E and F have a very favorable specificity toward NAG-NAM acceptors ($K_s \sim 3.5 \times 10^{-4}M$). Moreover, assuming[118] an association rate constant for reversible binding of an NAG-NAM moiety at EF of 10^6 to $10^8 M^{-1} \sec^{-1}$, this k_T/k_H ratio implies a lifetime for the glycosyl-enzyme intermediate in water of 10^{-3} to 10^{-5} second.

Experiments designed to test the hypotheses advanced on the basis of X-ray model building concerning catalysis by Glu 35 and Asp 52 and distortion in subsite D have been carried out in several laboratories. Using a carbodiimide reagent developed for selective modification of carboxyl groups in proteins, Koshland was able to demonstrate the absolute essentiality of Asp 52 to lysozyme catalysis.[119] Although the carboxyl group of Glu 35 proved totally unreactive to his reagent, all of the other carboxylic acid groups of lysozyme could be modified, and with the exception of Asp 52, they were all found to be nonessential to catalysis. Bruice has conducted hydrolysis studies on model acetals such as **XIII**,[111] **XIV**,[112] and **XV**[111,113] in attempts to shed light on the possible significance of neighboring-group carboxyl and carboxylate catalysis and ground-state planarity of cyclic acetal systems. The

---

* This stereochemistry cannot be determined when water is the acceptor because the product undergoes rapid mutarotation at the C_1-terminus.

XIII **XIV** **XV**

main conclusions from this work that bear on the lysozyme problem are, first, as stated earlier in a footnote, acid catalysis in all cases, including neighboring-group systems like **XIII** and **XV**, is always *specific* rather than *general*. That is, protonation of the leaving-group oxygen is complete before the transition state for bond rupture is reached. Second, in systems such as **XIII**, where neighboring-group electrostatic catalysis by a carboxylate ion might be expected, it is *not* observed. Of course the model systems were studied in water, and it is not known to what extent the possibility that lysozyme catalysis occurs in a different kind of microscopic environment might influence the catalytic behavior of Glu 35 and Asp 52 in the enzyme. Finally, compound **XIV**, in which the C_2, C_1, O, C_5 system is planar in the ground state, might be considered a model for the distorted pyranose ring occupying subsite *D* of lysozyme. This compound does indeed hydrolyze more than 10^2 faster than corresponding normal glycosides which assume the chair conformation. This rate enhancement is within the range of 10^2 to 10^4 cited earlier for the distortion effect as based upon the estimated binding free energy associated with subsite *D*.

In conclusion, it appears that the mechanistic model proposed for lysozyme action on the basis of crystallographic studies remains viable, although the model-systems studies of Bruice imply some uncertainty concerning the catalytic roles of Glu 35 and Asp 52 as an acid catalyst and an electrostatic catalyst, respectively.

References

1. M. Kunitz and J. H. Northrop, *J. Gen. Physiol.*, **19**, 991 (1936).
2. P. E. Wilcox, *Methods of Enzymology*, Vol. 19, Academic, New York, 1970, pp. 64 ff.
3. H. Neurath and G. Schwert, *Chem. Rev.*, **46**, 69 (1950).
4. M. L. Bender, F. J. Kezdy, and C. R. Gunther, *J. Am. Chem. Soc.*, **86**, 3714 (1964).
5. F. J. Kezdy and E. T. Kaiser, *Methods of Enzymology*, Vol. 19, Academic, New York, 1970, p. 1.
6. M. L. Bender and F. J. Kezdy, *Ann. Rev. Biochem.*, **34**, 49 (1965).

7. L. Cunningham in M. Florkin and E. H. Stotz (Eds.), *Comprehensive Biochemistry*, Vol. 16, Elsevier, Amsterdam, 1965, pp. 96–109.

8. G. P. Hess, in *The Enzymes*, 3rd ed., Vol. III, P. D. Boyer (Ed.), Academic, New York, 1971, pp. 213–248.

9. B. S. Hartley and B. A. Kilby, *Biochem. J.*, **50**, 672 (1952); **56**, 288 (1954).

10. A. K. Balls and H. N. Wood, *J. Biol. Chem.*, **219**, 245 (1956).

11. A. K. Balls, C. E. McDonald, and A. S. Bracher, *Proc. Intern. Symp. Enzyme Chem.*, *Tokyo*, *Kyoto*, Maruzen, Tokyo, 1958, p. 245.

12. G. H. Dixon, W. J. Dreyer, and H. Neurath, *J. Am. Chem. Soc.*, **78**, 4810 (1956).

13. E. F. Jansen, M. D. F. Nuttig, R. Jang, and A. K. Balls, *J. Biol. Chem.*, **179**, 189 (1949); A. K. Balls and E. F. Jansen, *Adv. Enzymol.*, **13**, 321 (1952).

14. D. E. Fahrney and A. M. Gold. *J. Am. Chem. Soc.*, **85**, 997 (1963).

15. D. E. Koshland, Jr., *Adv. Enzymol.*, **22**, 45 (1960).

16. W. Cohen, M. Lache, and B. F. Erlanger, *Biochemistry*, **1**, 686 (1962).

17. J. A. Cohen, R. A. Oosterbaan, H. S. Jansz, and F. Berends, *J. Cellular Comp. Physiol.*, **54** (Suppl. I), 231 (1959).

18. R. A. Oosterbaan, M. van Adrichem, and J. A. Cohen, *Biochem. Biophys. Acta*, **63**, 204 (1964).

19. B. S. Hartley, *Nature*, **201**, 1284 (1964); B. S. Hartley and D. L. Kauffmen, *Biochem. J.*, **101**, 229 (1966).

20. R. A. Oosterbaan and J. A. Cohen, in *Structure and Activity of Enzymes*, T. W. Goodwin, J. I. Harris, and B. S. Hartley (Eds.), Academic, New York, 1965, p. 87.

21. B. S. Hartley, J. R. Brown, D. L. Kauffmen, and L. B. Smillie, *Nature*, **207**, 1157 (1965).

22. B. S. Hartley, *Ann. Rev. Biochem.*, **29**, 45 (1960).

23. S. A. Bernhard, S. J. Lau, and H. Noller, *Biochemistry*, **4**, 1108 (1965).

24. S. A. Bernhard and H. Gutfreund, *Phil. Trans. Roy. Soc. London* (*B*), **257**, 105 (1970).

25. G. P. Hess, J. McCann, E. Ku, and G. McConkey, *Phil. Trans. Roy. Soc. London* (*B*), **257**, 89 (1970).

26. K. G. Brandt, A. Himoe, and G. P. Hess, *J. Biol. Chem.*, **242**, 3973 (1967).

27. D. M. Blow, in *The Enzymes*, 3rd ed., Vol. 3, P. D. Boyer (Ed.), Academic, New York, 1971, Chap. 6.

28. M. Bergmann and J. S. Fruton, *J. Biol. Chem.*, **118**, 405 (1938).

29. S. G. Cohen, V. M. Vaidya, and R. M. Schultz, *Proc. Nat. Acad. Sci. U.S.*, **66**, 249 (1970); S. G. Cohen, *Trans. N.Y. Acad. Sci.*, *Ser II*, **31**, 705 (1969).

30. G. E. Hein and C. Niemann, *J. Am. Chem. Soc.*, **84**, 4495 (1962).

31. T. A. Steitz, R. Henderson, and D. M. Blow, *J. Mol. Biol.*, **46**, 337 (1969).

32. A. J. Hymes, D. A. Robinson, and W. J. Canaday, *J. Biol. Chem.*, **240**, 134 (1965).

33. W. P. Jencks, *Catalysis in Chemistry and Enzymology*, McGraw-Hill, New York, 1969, Chap. 8.

34. S. G. Cohen, V. M. Vaidya, and R. M. Schultz, *Proc. Nat. Acad. Sci. U.S.*, **66**, 249 (1970).

35. D. W. Ingles and J. R. Knowles, *Biochem. J.*, **108**, 561 (1968); **104**, 369 (1967).

36. M. L. Bender, F. J. Kezdy, and C. R. Gunter, *J. Am. Chem. Soc.*, **86**, 3714 (1964).

37. J. T. Gerig, *J. Am. Chem. Soc.*, **90**, 2681 (1968).

38. G. P. Hess, in *The Enzymes*, 3rd ed., Vol. 3, P. D. Boyer (Ed.), Academic, New York, 1971, Chap. 7.

39. B. R. Hammond and H. Gutfreund, *Biochem. J.*, **61**, 187 (1955).

40. E. B. Ong, E. Shaw, and G. Schuellmann, *J. Am. Chem. Soc.*, **86**, 1271 (1964).

41. D. Glick, *Biochemistry*, **7**, 3391 (1968).

42. F. J. Kezdy, A. Thomson, and M. L. Bender, *J. Am. Chem. Soc.*, **89**, 1004 (1967).

43. E. Shaw, in *The Enzymes*, 3rd ed., Vol. 3, P. D. Boyer (Ed.), Academic, New York, 1970, Chap. 2.

44. Y. Nakagawa and M. L. Bender, *Biochemistry*, **9**, 259 (1970).

45. B. Labouesse, H. Oppenheimer and G. P. Hess, *Symposium on Structure and Activity of Enzymes*, T. W. Goodwin, J. I. Harris, and B. S. Hartley (Eds.), Academic, New York, 1964, p. 134.

46. H. Oppenheimer, B. Labouesse, and G. P. Hess, *J. Biol. Chem.*, **241**, 2720 (1966).

47. A. Himoe, P. C. Parks, and G. P. Hess, *J. Biol. Chem.*, **242**, 919 (1967).

48. J. Kraut, in *The Enzymes*, 3rd ed., Vol. 3, P. D. Boyer (Ed.), Academic, New York, 1971, p. 547.

49. J. H. Wang, *Science*, **161**, 328 (1968).

50. D. M. Blow, J. J. Birktoft, and B. S. Hartley, *Nature*, **221**, 337 (1969).

51. M. Caplow, *J. Am. Chem. Soc.*, **91**, 3639 (1969).

52. L. Polgar and M. L. Bender, *Proc. Nat. Acad. Sci. U.S.*, **64**, 1335 (1969).

53. A. R. Fersht and Y. Requena, *J. Am. Chem. Soc.*, **93**, 7079 (1971).

54. A. R. Fersht, *J. Am. Chem. Soc.*, **94**, 293 (1972).

55. J. March, *Advanced Organic Chemistry: Reactions, Mechanisms, and Structure*, McGraw-Hill, New York, 1968, p. 274.

56. M. H. O'Leary and M. D. Kluetz, *J. Am. Chem. Soc.*, **94**, 3585 (1972).

57. S. A. Bernhard, B. F. Lee, and Z. H. Tashjian, *J. Mol. Biol.*, **18**, 405 (1966).

58. M. L. Anson, *J. Gen. Physiol.*, **20**, 663 (1937).

59. H. Neurath, R. A. Bradshaw, P. H. Petra, and K. A. Walsh, *Phil. Trans. Roy. Soc. London (B)*, **257**, 159 (1970).

60. R. A. Bradshaw, L. H. Ericsson, K. A. Walsh, and H. Neurath, *Proc. Nat. Acad. Sci. U.S.*, **63**, 1389 (1969).

61. H. Neurath and G. W. Schwert, *Chem. Revs.*, **46**, 69 (1950).

62. H. Neurath, in *The Enzymes*, 2nd ed., Vol. 4, P. D. Boyer, H. A. Lardy, and K. Myrback (Eds.), Academic, New York, 1960.

63. J. A. Hartsuck and W. N. Lipscomb, in *The Enzymes*, 3rd ed., Vol. 3, P. D. Boyer (Ed.), Academic, New York, 1971.
64. I. Schechter and A. Berger, *Biochemistry*, **5**, 3371 (1966).
65. N. Abramowitz, I. Schechter, and A. Berger, *Biochem. Biophys. Res. Commun.*, **29**, 862 (1967).
66. B. L. Vallee, J. F. Riordan, J. L. Bethune, T. L. Coombs, D. S. Auld, and M. Sokolovsky, *Biochemistry*, **7**, 3547 (1968).
67. D. S. Auld and B. L. Vallee, *Biochemistry*, **10**, 2892 (1971).
68. B. L. Vallee, J. F. Riordan, D. S. Auld, and S. A. Lott, *Phil. Trans. Roy. Soc. London* (*B*), **257**, 215 (1970).
69. F. W. Carson and E. T. Kaiser, *J. Am. Chem. Soc.*, **88**, 1212 (1966).
70. J. E. Coleman and B. L. Vallee, *J. Biol. Chem.*, **236**, 2244 (1961).
71. R. T. Simpson, J. F. Riordan, and B. L. Vallee, *Biochemistry*, **2**, 616 (1963).
72. J. F. Riordan and B. L. Vallee, *Biochemistry*, **2**, 1460 (1963).
73. R. T. Simpson and B. L. Vallee, *Biochemistry*, **5**, 1760 (1966); O. A. Roholt and D. Pressman, *Proc. Nat. Acad. Sci. U.S.*, **58**, 280 (1967).
74. J. F. Riordan, M. Sokolovsky and B. L. Vallee, *Biochemistry*, **6**, 358 (1967).
75. M. Sokolovsky and B. L. Vallee, *Biochemistry*, **6**, 700 (1967).
76. B. L. Vallee and J. F. Riordan, *Brookhaven Symp. Biol.*, **21**, 91 (1968).
77. J. F. Riordan and H. Hagashida, *Biochem. Biophys. Res. Commun.*, **41**, 122 (1970).
78. P. H. Petra, *Biochemistry*, **10**, 3163 (1971).
79. G. M. Hass and H. Neurath, *Biochemistry*, **10**, 3535, 3541 (1971).
80. D. E. Koshland, Jr., *Proc. Nat. Acad. Sci. U.S.*, **44**, 98 (1958).
81. D. S. Auld and B. L. Vallee, *Biochemistry*, **9**, 602, 4359 (1970).
82. P. L. Hall, B. L. Kaiser, and E. T. Kaiser, *J. Am. Chem. Soc.*, **91**, 485 (1969).
83. W. N. Lipscomb, J. A. Hartsuck, G. N. Reeke, Jr., F. A. Quiocho, P. H. Bethge, M. L. Ludwig, T. A. Steitz, H. Muirhead, and J. C. Coppola, *Brookhaven Symp. Biol.*, **21**, 24 (1968).
84. J. Glovsky, P. L. Hall, and E. T. Kaiser, *Biochem. Biophys. Res. Commun.*, **47**, 244 (1972).
85. E. T. Kaiser and B. L. Kaiser, *Acc. Chem. Res.*, **5**, 219 (1972).
86. B. L. Kaiser and E. T. Kaiser, *Proc. Nat. Acad. Sci. U.S.*, **64**, 36 (1969).
87. B. L. Vallee, J. F. Riordan, J. L. Bethune, T. L. Coombs, D. S. Auld, and M. Sokolovsky, *Biochemistry*, **7**, 3547 (1968).
88. W. P. Jencks, *Catalysis in Chemistry and Enzymology*, McGraw-Hill, New York, 1969, Chap. 4.
89. L. M. Ginodman, N. I. Mal'tsev, and V. N. Orekhovich, *Biokimiya*, **31**, 1073 (1966).
90. P. L. Hall and E. T. Kaiser, *Biochem. Biophys. Res. Commun.*, **29**, 205 (1967).
91. G. Tomalin, B. L. Kaiser, and E. T. Kaiser, *J. Am. Chem. Soc.*, **92**, 6046 (1970).
92. A. Fleming, *Proc. Roy. Soc.* (*London*) (*B*), **93**, 306 (1922).
93. P. Jollès, *Angew. Chem. Int. Ed.*, **8**, 227 (1969).
94. E. P. Abraham and R. Robinson, *Nature*, **140**, 24 (1937).

95. P. Jollès, in *The Enzymes*, 2nd ed., Vol. 4, P. D. Boyer, H. Lardy, and K. Myrback (Eds.), Academic, New York, 1960, Chap. 25.

96. J. Jollès, J. Jauregui-Adell, I. Bernier, and P. Jollès, *Biochim. Biophys. Acta*, **78**, 668 (1963).

97. R. E. Canfield, *J. Biol. Chem.*, **238**, 2698 (1963).

98. C. C. F. Blake, D. F. Koenig, C. A. Mair, A. C. T. North, D. C. Phillips, and V. R. Sarma, *Nature*, **206**, 757 (1965).

99. L. R. Berger and R. S. Weiser, *Biochim. Biophys. Acta*, **26**, 517 (1957).

100. N. Sharon, T. Ozawa, H. M. Flowers, and R. W. Jeanloz, *J. Biol. Chem.*, **241**, 223 (1966).

101. C. C. F. Blake, G. A. Mair, A. C. T. North, D. C. Phillips, and V. Sarma, *Proc. Roy. Soc. (London) (B)*, **167**, 365 (1967).

102. L. N. Johnson and D. C. Phillips, *Nature*, **206**, 761 (1965).

103. D. C. Phillips, *Sci. Am.*, **215** (5), 78 (1966).

104. C. C. F. Blake, L. N. Johnson, G. A. Mair, A. C. T. North, D. C. Phillips, and V. R. Sarma, *Proc. Roy. Soc. (London) (B)*, **167**, 378 (1967).

105. D. C. Phillips, *Proc. Nat. Acad. Sci. U.S.*, **57**, 484 (1967).

106. R. E. Dickerson and I. Geis, *The Structure and Action of Proteins*, Harper and Row, New York, 1969.

107. D. Eisenberg, in *The Enzymes*, 3rd ed., Vol. 1, P. D. Boyer (Ed.), Academic, New York, 1970, Chap. 1.

108. S. M. Parsons and M. A. Raftery, *Biochemistry*, **11**, 1623 (1972).

109. F. W. Dahlquist and M. A. Raftery, *Biochemistry*, **9**, 3277 (1968).

110. C. A. Vernon, *Proc. Roy. Soc. (London) (B)*, **167**, 389 (1967).

111. B. Dunn and T. C. Bruice, *J. Am. Chem. Soc.*, **92**, 2410 (1970).

112. T. A. Giudici and T. C. Bruice, *Chem. Commun.*, **1970**, 690.

113. B. Dunn and T. C. Bruice, *J. Am. Chem. Soc.*, **93**, 5725 (1971).

114. D. M. Chipman and N. Sharon, *Science*, **165**, 454 (1969).

115. U. Zehari, J. J. Pollock, V. I. Teichberg, and N. Sharon, *Nature*, **219**, 1152 (1968).

116. M. A. Raftery and T. Rand-Meir, *Biochemistry*, **7**, 3281 (1968).

117. D. M. Chipman, *Biochemistry*, **10**, 1714 (1971).

118. G. G. Hammes, *Adv. Protein Chem.*, **23**, 1 (1968).

119. T. V. Lin and D. E. Koshland, Jr., *J. Biol. Chem.*, **244**, 505 (1969).

CHAPTER 10

Enzymic Oxidations

Oxidoreductases constitute one of the six major classes of enzymes (see Appendix). They are intimately involved in the utilization of the oxygen we breathe, taking part in a wide variety of biochemical processes. The study of many of these enzymes has been hampered by a combination of factors, including difficulty of isolation and characterization and, often, complicated kinetic behavior. However, because of their importance, a voluminous literature has been built up around them and a fairly organized pattern has emerged. Many of these enzymes require a metal ion cofactor and some utilize an organic coenzyme as well. These coenzymes frequently are oxidized by one enzyme of a chain of electron transfer enzymes and then the oxidized form will serve as the substrate for the next enzyme in the sequence or it will serve as a coenzyme to be reduced by that next enzyme, and vice versa. This regenerates the original coenzyme for use by the original enzyme of the sequence.

Classification

Basically these enzymes fall into two general classes, (1) substrate:O_2 oxidoreductases and (2) substrate:acceptor oxidoreductases. In the first class the substrate usually transfers its reducing equivalents to an intermediate carrier (e.g., $NAD^{\oplus}$), which then reacts directly with oxygen, yielding water and oxidized substrate. In the second class the substrate transfers its reducing equivalents to an intermediate carrier as before but now the carrier reacts with an acceptor other than O_2; this acceptor is usually then utilized as a substrate for another enzyme as mentioned above.

The overall oxidation reactions that occur in these systems generally involve either two or four electrons and one or two protons being transferred. These are typified by the following generalized reactions:

$$AH_2 + \tfrac{1}{2}O_2 \xrightarrow{\text{enzyme}} AH + 2H_2O_2 \qquad \text{(2-equivalent reduction)} \qquad \text{(10-1a)}$$

$$AH_2 + \tfrac{1}{2}O_2 \xrightarrow{\text{enzyme}} A + H_2O \qquad \text{(4-equivalent reduction)} \qquad \text{(10-2a)}$$

Similar general equations may be written for systems in which oxygen is not the acceptor:

$$AH_2 + R \xrightarrow{\text{enzyme}} AH + RH \qquad \text{(2-equivalent reduction)} \qquad \text{(10-1b)}$$

$$AH_2 + R \xrightarrow{\text{enzyme}} A + RH_2 \qquad \text{(4-equivalent reduction)} \qquad \text{(10-2b)}$$

It appears that a four-electron transfer reaction may involve intermediate generation of a hydroperoxide derivative of the enzyme upon oxidation of the first substrate molecule and that the subsequent reaction of this species with a second substrate molecule leads to the production of the two molecules of H_2O (see below).

In addition to these general classifications (and independent of them), a more detailed classification scheme has evolved for the enzymes catalyzing biological oxidations. More comprehensive treatises[1,2] discuss all of the classes of oxidoreductases; for our purposes, we need briefly describe only four major classes of oxidases: oxygenases, hydroxylases, dehydrogenases, and heme protein hydroperoxidases. The enzymes of these classes catalyze an enormous number of important reactions, including some of the enzymes of amino acid metabolism (e.g., transaminases, amine oxidases), carbohydrate metabolism (hexose oxidase, etc.) and lipid metabolism (lipoxygenase, acyl-CoA dehydrogenases, etc.).

a. Oxygenases

These enzymes, which are also referred to as either dioxygenases or oxygen transferases, cause the introduction of both atoms of a single oxygen molecule into a single substrate molecule (Eq. 10-3), producing a product that is for-

$$AH_2 + O_2^* \longrightarrow A(O^*H)_2 \qquad \text{(10-3)}$$

mally a dihydroxy derivative of the substrate. An example of this type of enzyme is pyrocatechase, whose reaction is shown in Eq. 10-4.

$$(10\text{-}4)$$

Another example in which the product is not actually a dihydroxy derivative of the substrate is the enzyme lipoxygenase.[3] This interesting enzyme, which apparently requires no cofactor of any kind, catalyzes the reaction of Eq. 10-5, which is the transformation of a *cis,cis*-1,4-pentadiene unit (usually in a linoleate chain) to a *cis,trans*-1,3-pentadiene structure with a hydroperoxide group inserted for one of the allylic protons at the 5 position.

$$R \underset{H}{\overset{1}{C}} = \overset{\overset{3}{CH_2}}{\underset{H}{C}} \underset{H}{\overset{5}{C}} = \overset{CH_2 - R'}{\underset{H}{C}} + O_2 \xrightarrow{\text{lipoxygenase}}$$

(10-5)

$$\overset{OOH}{\underset{R}{\overset{H}{\underset{1}{C}}} = \overset{3}{\underset{H}{C}} \overset{5}{\underset{H}{C}} H - CH_2R'}$$

b. Hydroxylases

These enzymes are commonly referred to as monooxygenases or mixed-function oxidases. If one studies the reactions of these enzymes using iso-topically labeled oxygen, the label distribution in the products is as illustrated in Eq. 10-6. Enzymes of this class have an absolute requirement for a second

$$AH_2 + BH_2 + O_2^* \xrightarrow{\text{hydroxylase}} AHO^*H + HO^*H + B \qquad (10\text{-}6)$$

oxidizable substrate (BH_2) and this helps distinguish them from the oxygen-ases mentioned previously. A typical reaction for this class of enzymes is shown in Eq. 10-7 for the enzyme phenolase, which catalyzes the oxidation

$$+ H_2O^* \quad (10\text{-}7)$$

of a substituted phenol to the corresponding catechol. The required cosub-strate is a substituted catechol, which is transformed to the quinone. The elegant work of Witkop and Daly and their co-workers[4] at the National Institutes of Health has shown that in very many cases the substituent at the site of oxygen substitution in the substrate (labeled T) migrates to the ad-jacent position in the molecule during the reaction (NIH shift), and this mechanistic feature must be borne in mind when studying these enzymes.

c. Dehydrogenases

According to Eqs. 10-1 and 10-2, the reactions presented thus far have been four-equivalent redox reactions; that is, each atom of oxygen from O_2 gains two electrons (four total). The dehydrogenases can participate in a number of four-equivalent redox processes and also provide a large number of examples of two-equivalent processes. Thus the laccase (polyphenol oxidase) process is a four-equivalent process (Eq. 10-8), as is the ascorbic acid oxidase process (Eq. 10-9), whereas the diamine oxidase process (Eq. 10-10) is a

$$2 \quad \text{[structure: } p\text{-dihydroxybenzene]} + O_2 \xrightarrow{\text{laccase}} 2 \quad \text{[structure: } p\text{-benzoquinone]} + 2H_2O \tag{10-8}$$

$$2 \quad \text{[structure: ascorbic acid]} + O_2 \xrightarrow{\text{AAO}} \quad \text{[structure: dehydroascorbic acid]} + 2H_2O \tag{10-9}$$

two-equivalent process, producing H_2O_2. Formally, each atom of from O_2 in Eq. 10-10 gains one electron (two total).

$$RCH_2NH_2 + O_2 + H_2O \xrightarrow[\text{oxidase}]{\text{diamine}} RCHO + NH_3 + H_2O_2 \tag{10-10}$$

Later in this chapter we shall see that conclusions about mechanisms on the basis of whether a system is a two- or four-equivalent process are tenuous at best when we examine two examples of these dehydrogenases.

d. Heme Protein Hydroperoxidases

Amine oxidases as exemplified by Eq. 10-10 are common in biological systems and in order to prevent the buildup of the toxic product hydrogen peroxide, the heme protein hydroperoxidases have evolved. There are two classes of these enzymes, the catalases and the peroxidases. Catalases decompose H_2O_2 to O_2 and H_2O (the catalatic reaction; Eq. 10-11) in the absence of

$$2H_2O_2 \xrightarrow{\text{catalase}} 2H_2O + O_2 \tag{10-11}$$

any exogenous hydrogen donor. This is one of the fastest enzymic reactions yet measured, with a turnover number of about 10^7 to $10^8 M/(\text{liter})(\text{sec})$ per mole of enzyme. Peroxidases catalyze the reaction of H_2O_2 with a wide variety of exogenous hydrogen donors including hydroquinone (Eq. 10-12).

$$\text{[structure: hydroquinone]} + H_2O_2 \xrightarrow{\text{peroxidase}} \text{[structure: } p\text{-benzoquinone]} + 2H_2O \tag{10-12}$$

Whether peroxidase has any function other than peroxide decomposition (a detoxification function) is still being debated. It has been implicated in melanin formation (the basic skin and hair pigment). For more information on these useful enzymes, the reader is referred to more comprehensive references.[5]

The Hamilton Theory of Mechanism in Biological Oxidation Reactions

Our understanding of mechanisms of biological redox reactions is significantly less advanced than our understanding of the mechanisms of hydrolytic enzymes. In large measure this stems from the nonavailability of suitable non-enzymic model systems that allow detailed studies of a reaction mechanism to be made. As indicated in Chapter 9, an important feature in the mechanisms of hydrolytic enzymes is the general acid or base catalyzed proton transfer step. In view of the fact that biological oxidations take place in essentially the same bulk environment as do the nonredox reactions, one might ask the question "Are general acid or base catalyzed proton transfers crucially involved in the mechanisms of the redox enzymes?" Hamilton has raised this question and presented some cogent arguments that proton transfers are indeed involved in the mechanisms of many oxidative enzymes, and importantly so.[6] This hypothesis is then used to propose for a host of biological oxidations mechanisms that have as a unifying feature the transfer of all hydrogen atoms as protons; the electrons transfer either by moving through the network of conjugation provided by the substrate or a substrate-coenzyme intermediate or through a metal ion-substrate or metal ion-substrate-oxygen intermediate. The existence of free radicals is suggested to occur only when the radical formed is very stable, as in the hydroquinone to semiquinone radical transformation.

Let us now examine this theory in greater detail. As mentioned earlier, biological oxidations are either two or four-electron transfer processes involving organic compounds. Unlike metal ions, such as $Fe^{2+} \rightarrow Fe^{3+}$, the "valence" of carbon does not change upon oxidation or reduction because all electrons are paired in both the starting materials and the product. Thus in organic oxidations, electrons are transferred in multiples of two. For example, in the oxidation of methanol to formaldehyde (Eq. 10-13), the

$$\text{H} \overset{\text{H}}{\underset{\text{H}}{:\ddot{C}:\ddot{O}:H}} \xrightarrow[\text{[O]}]{} \text{H}\overset{\text{H}}{:\ddot{C}::\ddot{O}} + 2H^{\oplus} + 2e^{\ominus} \tag{10-13}$$

carbon atom maintains four bonding electrons in both compounds but the total system loses two electrons (and the carbon atom hybridization changes from sp^3 to sp^2). This has two important consequences in biochemistry. First it implies that radicals, generated by one-electron transfer, will occur only with great difficulty or when the radical formed is a very stable one. Second, it means that most organic compounds will not react readily with oxygen under physiological conditions, because O_2 is a triplet molecule and most organic compounds are singlets (all electrons paired), causing any un-catalyzed organic oxidation to be a spin forbidden process. Nature has evolved the use of metal ions to help catalyze these reactions where needed.

The process of Eq. 10-13 formally involves the loss of a hydrogen molecule (a two-electron process); theoretically, this could occur through the involve-ment of protons (P), hydrogen atoms (A) or hydride ions (H_y). This gives rise to six possible mechanisms for the dehydrogenation:

1. The AA mechanism ($H \cdot + H \cdot$): Both hydrogens transferred as hydro-gen atoms (common in organic chemistry).
2. The AH_y mechanism ($H \cdot + H:^{\ominus} - 1e^{\ominus}$): One hydrogen transferred as a hydrogen atom, one as a hydride ion, and one electron by another path-way.
3. The AP mechanism ($H \cdot + H^{\oplus} + 1e^{\ominus}$): One hydrogen transferred as a hydrogen atom, one as a proton, and one electron by another pathway.
4. The H_yH_y mechanism ($H:^{\ominus} + H:^{\ominus} - 2e^{\ominus}$): Both hydrogens transferred as hydride ions and two electrons by another pathway.
5. The PH_y mechanism ($H^{\oplus} + H:^{\ominus}$): One hydrogen transferred as a proton, one as a hydride ion.
6. The PP mechanism ($H^{\oplus} + H^{\oplus} + 2e^{\ominus}$): Both hydrogens transferred as protons and two electrons by another pathway.

In all these mechanisms electrical neutrality is maintained by transfer of the appropriate number of electrons in the proper direction.

Hamilton proposes that the PP mechanism is particularly attractive for biological oxidations because the hydrogens are transferred as protons and thus could be susceptible to normal types of acid and base catalysis so preva-lent in the mechanisms of nonoxidative enzymes. In addition to this intuitively appealing rationale, there are convincing theoretical arguments as to why hydrogen should transfer as a proton in biological systems.

As mentioned previously, the organic radicals derived from most organic biological compounds are unstable. Since these would have to form from the loss of a hydrogen atom, the transfer of hydrogen as $H \cdot$ is therefore highly unlikely. In addition, hydrogen is bonded to atoms of equivalent or greater electronegativity, making it difficult at best to generate a hydride ion. Thus at 37°C and pH values between 6 and 9, hydrogen atom or hydride ion transfers are expected to be slow processes. Proton transfers, however, are known to be remarkably facile under these conditions. In fact, the rates of proton transfer to or from nitrogen, oxygen, or sulfur have been shown by Eigen to be diffusion controlled.[7] Proton transfer reactions to carbon are somewhat slower but are still rapid if they lead to stable products. Physically, this much more favorable rate of proton transfer (compared to $H\cdot$ or $H:^{\ominus}$ transfer) might be explained by the much smaller mass/volume ratio for protons as opposed to any species containing both protons and electrons ($H\cdot$ or $H:^{\ominus}$ have the same mass as $H^{\oplus}$ but significantly greater particle volume). Additionally, a proton is able to penetrate the electron cloud of an atom it is approaching much more readily than a species that has its own electron cloud associated with it. In fact, coulombic forces should be favorable until the proton approaches the receiving atom's nucleus.*

Given the foregoing reasoning, how then does the PP mechanism manifest itself? First, let us consider a simple PP process, the intramolecular alcohol oxidation–ketone reduction of Eq. 10-14. This nonenzymic reaction occurs

$$\text{(10-14)}$$

readily with either acid or base catalysis. The electrons transfer through a short π-electron system and hydrogens transfer as protons. Few biological systems are this simple.

Consider now the more complex process of Eq. 10-15. This is another

$$\text{(10-15)}$$

intramolecular oxidation-reduction reaction that could occur by $H^{\oplus}$ and electron transfer, as shown. However, most biological redox reactions are not intramolecular and therefore the extended π-system does not, a priori, exist. We may ask, for example, how Eq. 10-15 proceeds as an intermolecular reaction in biological systems. Hamilton[6] proposes three variations of the

---

* Although qualitatively appealing, these arguments are admittedly oversimplifications since the species involved would be hydrated in aqueous media. The nature of solvation at an enzyme active site is unknown, however.

PP mechanism which not only help rationalize this situation but quite likely rationalize a wide variety of enzymic oxidation reactions: the PPR mechanism the PPC mechanism, and the PPM mechanism.

a. The PPR Mechanism

This mechanism involves radical (R) intermediates, but as before, hydrogens transfer as protons and electrons as electrons. Intermolecularly, Eq. 10-15 appears as Eq. 10-16. Equation 10-17 depicts how this might occur by the

$$\text{(catechol)} + R\text{—}N{=}N\text{—}R \rightleftharpoons \text{(ortho-quinone)} + RNH\text{—}NHR \qquad (10\text{-}16)$$

$$(10\text{-}17)$$

PPR mechanism. Clearly, a process such as Eq. 10-17 would not be expected to occur unless the radical species generated are quite stable, as is the case for semiquinone radicals. The stability of the nitrogen radicals shown in Eq. 10-17 would depend on the nature of R.

b. The PPC Mechanism

Here again hydrogens transfer as protons and electrons as electrons, but in addition the process depends upon the formation of a covalent intermediate, hence the designation PPC. Thus the process of Eq. 10-16 occurring by the PPC mechanism would appear as in Eq. 10-18. Note that no radical species

$$(10\text{-}18)$$

are generated, and reactions of this sort might be considered "quasi-intra-molecular," since the second step is indeed unimolecular. As alluded to in Chapter 8, this mechanism is particularly appealing as a rationale for enzymes requiring certain coenzymes, such as $NAD^{\oplus}$, pyridoxal phosphate, or the flavins. These three coenzymes will be considered again at the end of this chapter.

c. The PPM Mechanism

As in the PPR and PPC cases, this mechanism also transfers hydrogens as protons and electrons as electrons. The distinguishing feature here is the involvement of a metal ion (M of PPM). In these processes, no radical species are involved; rather, the function of the metal ion is to link two reacting species in such a way as to provide a greatly extended π-system, which allows all electron shifting to occur in what are essentially intramolecular processes. The ability of the first row of transition elements to facilitate rapid electron transfer between two π-bonded ligands makes them particularly well suited for biochemical reactions utilizing the PPM mechanism and helps explain their common occurrences in biological redox systems (especially Fe, Cu, and Zn). The process of Eq. 10-16 occurring by the PPM mechanism would appear as Eq. 10-19.

$$(10\text{-}19)$$

In summary, the Hamilton proposal says that in general, the PP mechanism is the preferred mechanism for biological redox reactions, and in any particular systems, one of the three variations—PPR, PPC, or PPM—must occur (unless the reaction is completely intramolecular, as in Eq. 10-14). If experimental evidence indicates the presence of free radicals (EPR spectra, radical scavenging experiments, etc.), then the mechanism is most likely PPR. If not, then it is PPC or PPM. If no radicals are involved and no metal ions are found, then a covalent intermediate is required.

It should be emphasized here that these proposals, although appealing, are presently based mainly on qualitative reasoning. Much careful work on many oxidative enzymes will be required to substantiate them, for the theory is by no means universally accepted at this time.

Two Examples

We now turn briefly to two examples of four-equivalent oxidations catalyzed by enzymes classified as dehydrogenases. It will be seen that for one the experimental evidence suggests the PPR mechanism, while for the other, the PPM mechanism is preferred. This will be followed by a consideration of the PPC mechanism for the coenzymes $NAD^{\oplus}$, FAD, FMN, and PLP.

a. Ascorbic Acid Oxidase (EC 1.10.3.3)

Reaction 10-9 is the ascorbic acid oxidase (AAO) reaction. Note that it is a typical dehydrogenase reaction—a four-electron transfer dehydrogenation of ascorbic acid and concomitant reduction of O_2 to H_2O. H_2O_2 has never been observed in AAO reactions, although attempts have been made to detect it. This enzymic reaction was first observed to occur in 1928, and the enzyme was isolated* from squash in 1935.[8] It is a blue, copper-containing protein, having eight atoms of Cu per molecule.[9] Ultracentrifuge studies have established its molecular weight at 146,000 ($\pm$ 15,000); electrophoresis places the isoelectric point between pH 5.0 and 5.5. Below pH 4.0 the Cu dissociates from the enzyme, causing inactivation. Moreover, the enzyme is not particularly stable even above 4.0, showing a gradual loss of activity at 0–5°C at its pH optimum of 5.6. This pH 5.6 activity loss can be largely prevented by the addition of an inert protein (routinely gelatin), which seems to protect the AAO from a conformational unfolding.

There are apparently no titratable sulfhydryl (SH) groups in the native enzyme. However, in $8M$ urea or in $0.2M$ sodium dodecyl sulfate the Cu

---

* One must be careful in attempting to isolate "AAO activity" to ensure that activity peaks are proteinaceous since Cu^{2+} and a variety of other oxidative enzymes can oxidize ascorbic acid to dehydroascorbic acid, often, however, producing H_2O_2 instead of H_2O from the reacting oxygen.

is liberated and from 10 to 12 SH groups become titratable. Complete reduction of the copper-free protein with mercaptoethanol shows from 16-20 SH groups, indicating between three and five disulfide bonds per molecule in addition to the copper-bound sulfhydryl groups. If one adds cuprous ion [Cu(I)] to a solution of the apoenzyme, full activity is restored upon the addition of 7.5–8 atoms of Cu(I) per mole of enzyme.[9] Predictably, chelating agents capable of removing the copper will inactivate the enzyme.

Estimates of K_m for ascorbic acid range from $4 \times 10^{-5}M$ to $5 \times 10^{-3}M$, depending on the conditions used. Turnover numbers of about 4×10^5 moles per liter of ascorbic acid oxidized per minute per mole of enzyme have been reported.

In addition to L-ascorbic acid, we know that L-glucoascorbic acid, L-galacto-ascorbic acid, and other ascorbic acid derivatives are susceptible to enzymic oxidation. In one study of the enzyme's specificity at pH 7.2 it was found that hydroquinone and catechol may also be oxidized and it was proposed that any conjugated anion that can form a semiquinone intermediate and a quinoidal product is capable of being oxidized.[8]

The oxidation state of copper in this enzyme is not uniform; there are apparently both Cu(II) and Cu(I) species present. Moreover, there is some evidence[8] that a reversible reduction-oxidation of copper takes place during the enzymic process (see below).

In considering a molecular mechanism for this enzyme one must remember that in contrast to the reaction of free Cu(II) with ascorbic acid (Eq. 10-20),

$$O_2 + Cu(II) + \text{ascorbic acid anion} \longrightarrow$$

$$\text{dehydroascorbate} + H_2O_2 + Cu(I) \qquad (10\text{-}20)$$

$$\overset{O_2}{\longrightarrow} Cu(II)$$

where H_2O_2 is produced, the only reduction product of O_2 in the enzyme process is H_2O. A modification of a scheme presented by Hamilton[10] for the oxidation of ascorbic acid by copper chelates is presented in Eq. 10-21.

$$(10\text{-}21)$$

This process seems applicable to the enzymic system since it provides for the reduction of Cu(II) to Cu(I) and it also invokes the generation of the ascorbate radical, which Yamazaki and Piette[9] have shown by EPR to be produced by the enzyme during its oxidation of ascorbic acid. Their data imply that a Cu(II)-E system is reduced to a Cu(I)-E system when the radical is generated. This radical is released to solution and subsequently reacts with another molecule of ascorbate radical in a nonenzymic disproportionation reaction, as shown in Eq. 10-22. The enzymic portion of the reaction, as

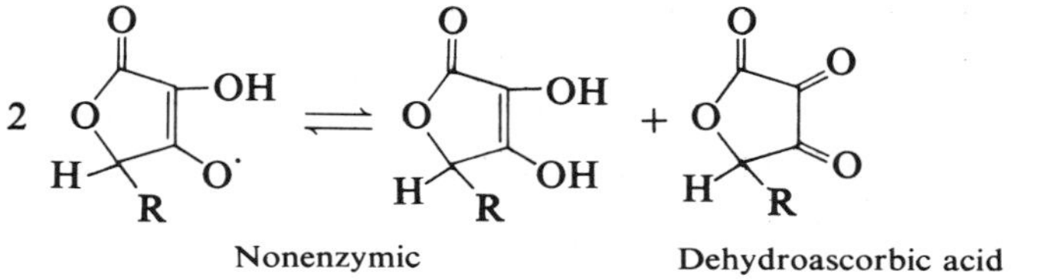

$$(10\text{-}22)$$

Nonenzymic   Dehydroascorbic acid

drawn, is an example of the PPR mechanism where hydrogen transfers as a proton and a fairly stable radical is generated.

As written in Eq. 10-21, the overall enzymic reaction is not complete, since, as mentioned earlier, in the ascorbate oxidase reaction a mole of oxygen is converted to two moles of H_2O. The PPR process of Eq. 10-21 results in the transfer of one electron to one of the eight Cu(II) atoms on the enzyme, yielding a Cu(I) atom. Since there is no evidence for the generation of an enzyme-bound radical, Hamilton[6] conceives of the regeneration of native Cu(II)-enzyme occurring as in Eq. 10-23 after first generating the quadruply reduced structure **I** by reacting with four ascorbate molecules. In order for this to occur the four Cu(I) atoms must somehow be electronically connected,

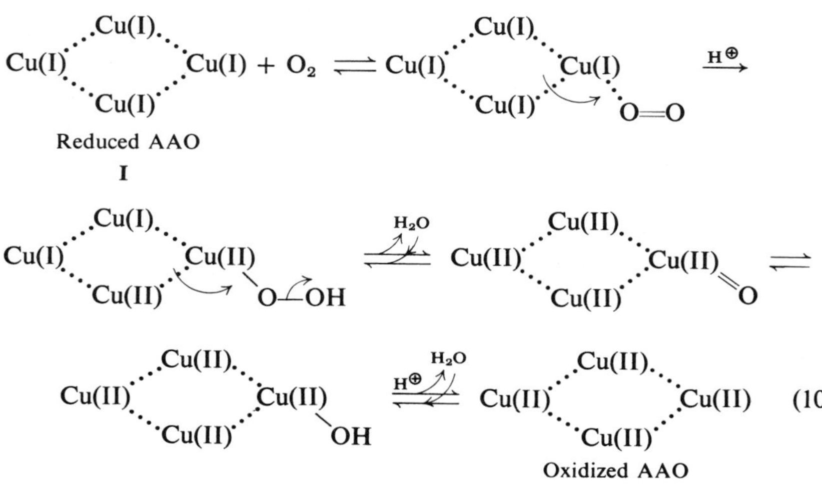

Reduced AAO

I

$$(10\text{-}23)$$

Oxidized AAO

since each Cu(I) transfers only one electron as it goes to Cu(II), whereas both of the steps in the reduction of O_2 are two-electron transfers. As drawn, Eq. 10-23 may be considered a PPM process. Thus the overall ascorbic acid oxidase reaction may utilize both the PPR and PPM processes.

b. Tyrosinase (EC 1.10.3.1)

Tyrosinase can catalyze two separate reactions, as exemplified by Eqs. 10-7 and 10-8. Reaction 10-7 is typical of the hydroxylase activity of this enzyme. This activity is of limited interest to our discussion and the reader is directed to the articles in reference 4 for a deeper discussion of it. The dehydrogenase type of activity of this enzyme, although similar in nature to Eq. 10-8, is restricted to *ortho*-substituted dihydroxybenzene derivatives, as shown in Eq. 10-24. Aside from this restriction, tyrosinase is quite non-specific, oxidizing a wide variety of catechol derivatives. Because of these

$$2 \quad \text{(catechol)} \quad + O_2 \xrightarrow{\text{tyrosinase}} 2 \quad \text{(ortho-quinone)} \quad + 2H_2O \qquad (10\text{-}24)$$

two distinct activities, much confusion exists concerning the nomenclature of this enzyme, with names such as catechol oxidase, dopa oxidase, phenolase complex, polyphenol oxidase, and tyrosinase all being frequently encountered. We use the name tyrosinase exclusively.

The richest known source of the enzyme is the mushroom,[12] which elaborates it to the extent of 40 μg per gram of plant tissue (0.004%). A typical isolation process involves first pressing the tissue with a hydraulic press to express the fluids, washing the pulp with H_2O, and then putting the combined liquids through an eight-step process involving (among other steps) ammonium sulfate precipitation, absorptions with calcium phosphate, and acetone precipitations.

The molecular weight of tyrosinase as determined from either light scattering or ultracentrifugal measurements is 130,000 ($\pm$ 10,000). It has an isoelectric point of about 5.0 and a copper content of about 0.2–0.3%, which implies approximately five copper atoms per molecule. In contrast to ascorbic acid oxidase, which has both Cu(II) and Cu(I), tyrosinase has copper only in the cuprous form, and its oxidation state does not change during catalysis. The copper may be removed by treatment with 0.01M HCN followed by exhaustive dialysis against neutral phosphate buffer. This leads to a complete loss of activity, which can be reversed by adding back an amount of Cu(I) equivalent to that removed. Many studies[12] with sulfhydryl reagents have led to the conclusion that there are no SH groups, even in the copper-free enzyme. Thus the mode of copper binding for tyrosinase must be different from that for ascorbic acid oxidase.

In the conversion of an *o*-dihydroxybenzene derivative to the *o*-quinone, tyrosinase acts as a four-electron transferase, converting 1 mole of oxygen to 2 moles of water during the oxidation of 2 moles of substrate. The quinoidal products are themselves reactive species and it is quite frequently observed that more than the stoichiometric quantity of O_2 is consumed due to further nonenzymic oxidation of these quinones.

Mason et al.[13] have studied the reaction of tyrosinase with catechol using electron paramagnetic resonance techniques and have concluded that there are no radical species produced by the enzymic process. Hamilton[10] has suggested that the PPM mechanism is attractive for this type of nonradical oxidation, involving the loss of two electrons and two protons from each substrate molecule in a stepwise fashion. This is shown in Eq. 10-25. As discussed earlier, this type of mechanism is appealing in that it allows the oxidation to proceed through a completely ionic pathway and it also allows the

$$2OH^{\ominus} + E \quad (10\text{-}25)$$

use of general acid and general base catalysis mechanisms to facilitate the proton and electron transfers. These processes were seen in Chapter 9 to be of extreme importance to the hydrolases. In essence, the Cu atom brings substrates catechol and oxygen together to form a pseudo-aromatic complex where electron transfer may occur readily, particularly when assisted by a general acid or base (as appropriate) from the protein.

This mechanism is open to question, however, and deeper study is called for. For example, if the reaction were run with equivalent molar quantities of O_2, catechol, and enzyme, followed by dialysis, species I might be isolable and should react with a second equivalent of catechol in the absence of oxygen. In fact, it should be emphasized that, appealing as it is, the Hamilton theory is just that, a theory, and will require much careful study in the future.

The two enzymes that have just been briefly discussed exemplify the complexity of the biological oxidation enzymes. On the surface, they appear sufficiently similar that one might expect them to utilize the same molecular mechanism. They are of similar molecular weight, have similar copper content, have similar isoelectric points, catalyze similar types of reaction (dehydrogenations)—both by four electron transfer processes (no H_2O_2 produced), and have no detectable sulfhydryl groups in their native structures. With all these similarities, however, further probing shows them to be significantly different in three important respects. First, whereas tyrosinase shows no free sulfhydryl groups in the copper-free enzyme, ascorbic acid oxidase shows many. This indicates fundamental differences in copper binding to the protein. Second, the copper of tyrosinase is all Cu(I), whereas the copper of ascorbic acid oxidase is about 50% Cu(I) and 50% Cu(II). Third—and very important—the mechanisms are dramatically different, with tyrosinase utilizing an ionic process (probably PPM) requiring no change of copper oxidation state during the catalytic process, whereas ascorbic acid oxidase utilizes a free radical mechanism (probably PPR), resulting in a change in copper oxidation state during the process from Cu(II) to Cu(I), and then back to Cu(II), probably via a PPM process.

The point to be learned here is that gross similarities seldom suffice to prove identity of catalytic mechanism; thus when studying an enzyme, there is a clear need for the sort of detailed examination that permits a choice of mechanism to be made. In fact, there is still much to be done on the two examples discussed here, since Eqs. 10-21 and 10-25 are merely plausible suggestions and since no clear function for the protein portion of either enzyme has been demonstrated.

Coenzymes and the PP Mechanism

The common coenzymes were considered in Chapter 8. The reactions catalyzed by enzymes utilizing many of these coenzymes can be considered organic redox reactions, and for such systems it seems probable that some variant of the PP mechanism will be appropriate. In this section, we will briefly consider this prospect for three coenzymes involved in an enormous number of biochemical processes: nicotinamide adenine dinucleotide ($NAD^{\oplus}$), flavin adenine dinucleotide (FMN or FAD), and pyridoxal phosphate (PLP). The total structures are shown in Table 8-1, p. 136; for our purposes it is sufficient to redraw the reactive ends of these molecules, with the only modification being that PLP is shown as the most likely active form in an enzymic reaction—an aldimine formed by reaction with an apoenzymic amino group. Thus we have structures **I, II,** and **III** for NAD⊕, FAD or FMN, and PLP, respectively.

In each of these figures a carbon atom has been marked with an asterisk. If we examine these * carbon atoms, we find that in each case it is an aldimine carbon atom. As mentioned in Chapter 8 in discussing PLP, an aldimine carbon atom is about 30 times more reactive to nucleophilic attack than is a carbonyl carbon atom. In addition, one might expect each of these aldimine carbon atoms to be considerably more reactive because each is significantly more electron deficient than a normal aromatic aldimine carbon atom would be. Thus the pyridine nitrogen of **I** is protonated, which should considerably enhance the susceptibility of C_2 to nucleophilic attack. (As mentioned in Chapter 8, this position is indeed the kinetically favored site of nucleophilic addition, although model system studies tend to favor C_4 as the position of attack.) The reactivity of the aldimine carbon of the flavins should be enhanced by the electron-withdrawing effects of the adjacent amide and amidine carbon atoms (1 and 2). The aldimine carbon atom of **III** should be activated toward nucleophilic attack by the electron-withdrawing effects of the substituted pyridine ring.

In view of these considerations, it seems clear that all three of these coenzymes are properly designed for a PPC mechanism. General examples of such mechanisms are shown in Eq. 10-26 for an NAD⊕-catalyzed dehydrogenation of an alcohol, Eq. 10-27 for a flavinoid-catalyzed dehydrogenation of an alcohol, and Eq. 10-28 for a PLP-catalyzed transamination. (Equations 10-26 and 10-28 have been presented previously in Chapter 8.) As required by the PPC mechanism, all three reactions involve the formation of a covalent intermediate and the transfer of all hydrogen atoms as protons with electron transfer occurring separately from proton transfer. As indicated in

$$(10\text{-}26)$$

$$(10\text{-}27)$$

Eq. 10-28, it is assumed that in all such enzymic reactions utilizing this mechanism general acids or bases will be suitably positioned on the enzyme to assist the proton transfers. (It should be pointed out that the proposed cyclic process of Eq. 10-26 makes this proton transfer process indistinguishable from a cyclic process involving a hydride ion transfer. If this is actually the case, then Eq. 10-26 would be an example of a "PHyC" mechanism.)

In closing, let it be said once again that the Hamilton proposal is a theory, and a relatively new one at that. As such it should be examined and tested very carefully in any system where one might wish to apply it. Its chief shortcoming is the scarcity of well-designed experiments to test it over the range of biological systems it includes. For example, there is much indirect evidence from model-systems studies suggesting that in $NAD^{\oplus}$ systems (Eq. 10-26), alcohol dehydrogenation occurs through hydride transfer directly and stereospecifically to C_4 of the nicotinamide ring.[14] Aside from the intuitive theoretical arguments discussed here, there is little experimental evidence supporting the proposal of Eq. 10-26. The chief appeal of the theory to the chemist is that it provides a mechanism whereby biological oxidations

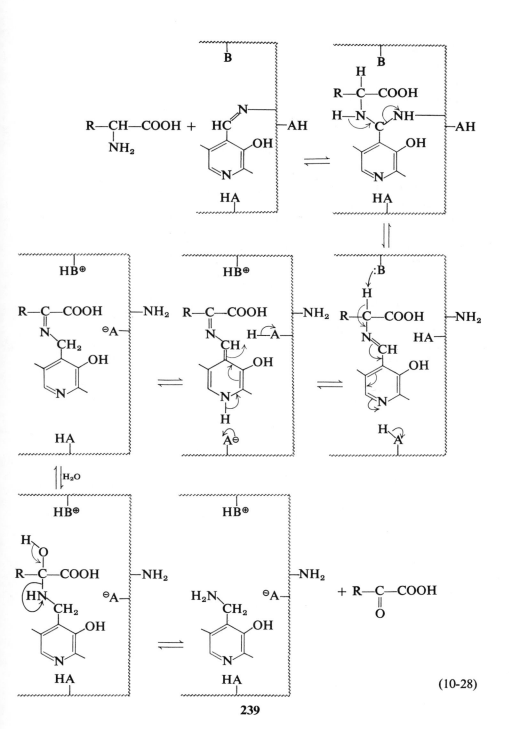

$$(10\text{-}28)$$

239

can begin to be understood in terms of the familiar physical organic chemical principles that have already achieved a moderate success in explaining some nonoxidative enzymic reactions.

References

1. T. P. Singer (Ed.), *Biological Oxidations*, Wiley, New York, 1966.
2. H. R. Mahler and E. H. Cordes, *Biological Chemistry*, Harper and Row, New York, 1966, Chap XIV.
3. F. C. Stevens, D. M. Brown, and E. L. Smith, *Arch. Biochem. Biophys.*, **136**, 413 (1970).
4. J. W. Daly, G. Guroff, D. M. Jerina, S. Udenfriend, and B. Witkop, *Adv. Chem. Ser.*, **77**, 279 (1968); D. M. Jerina, J. W. Daly, and B. Witkop, *The Role of Biogenic Amines and Physiological Membranes in Modern Drug Therapy*, Marcel Dekker, New York, 1970.
5. B. C. Saunders, A. G. Holmes-Siedle, and B. P. Stark, *Peroxidase*, Butterworth, London, 1964; P. D. Boyer, H. Lardy, K. Myrback (Eds.), *The Enzymes*, 2nd ed., Vol. 8, Academic, New York, 1963, Chaps, 6,7.
6. G. A. Hamilton, in *Progress in Bioorganic Chemistry*, E. T. Kaiser and F. J. Kezdy (Eds.), Wiley-Interscience, New York, 1971, pp. 83–157.
7. M. Eigen, *Angew. Chem. Int. Ed.*, **3**, 1 (1964).
8. G. R. Stark and C. R. Dawson, in *The Enzymes*, Vol. 8, P. D. Boyer, H. Lardy, and K. Myrback (Eds.), Academic, New York, 1963, Chap. 10.
9. Z. G. Penton and C. R. Dawson, in *Oxidases and Related Redox Systems*, T. E. King, H. S. Mason, and M. Morrison (Eds.), Wiley, New York, 1965, p. 222.
10. G. A. Hamilton, *Adv. Enzymol.*, **32**, 55 (1969).
11. I. Yamazaki and L. H. Piette, *Biochim. Biophys. Acta*, **50**, 62 (1961).
12. D. Kertesz and R. Zito, in *Oxygenases*, O. Hayaishi (Ed.), Academic, New York, 1962, p. 307.
13. H. S. Mason, E. Spencer, and I. Yamazaki, *Biochem. Biophys. Res. Commun.*, **4**, 236 (1961).
14. J. Everse, E. C. Zoll, L. Kahan, and N. O. Kaplan, *Bioorganic Chemistry*, **1**, 207 (1971); see also references cited therein.

CHAPTER 11

Fructose Diphosphate Aldolase

Fructose diphosphate aldolase (FDP-A; EC 4.1.2.13) is classified as a lyase. "A lyase is an enzyme which can remove groups from their substrates (not by hydrolysis) leaving double bonds, or which conversely adds groups to double bonds."[1] It is one of the more widely studied examples of this general enzyme class and is of interest to us for that reason and also because it has been found that there are two general types of FDP-aldolases that have evolved independently and that utilize distinctly different chemical approaches to catalyze the same chemical reaction—the reversible cleavage of fructose diphosphate. It is an important type of glycolytic enzyme very widely distributed in nature, being found in most plant and animal tissues and in many microbiological systems as well.

The reactions of interest to us that are catalyzed most readily by FDP-A are shown in Eqs. 11-1 and 11-2.[2] These are the cleavages of fructose-1,6-diphosphate (FDP) (Eq. 11-1) and the cleavage of fructose-1-phosphate (F-1-P) (Eq. 11-2).

$$
\begin{array}{c}
\underset{\text{(FDP)}}{\begin{array}{c}
\text{CH}_2\text{OP}\overset{\ominus}{\text{O}}_3\text{H} \\
| \\
\text{C}{=}\text{O} \\
| \\
\text{HOCH} \\
| \\
\text{CHOH} \\
| \\
\text{CHOH} \\
| \\
\text{CH}_2\text{OP}\overset{\ominus}{\text{O}}_3\text{H}
\end{array}}
\quad
\underset{\Longleftrightarrow}{\overset{\textbf{FDP-A}}{}}
\quad
\begin{array}{c}
\underset{\text{(DHAP)}}{\begin{array}{c}
\text{CH}_2\text{OP}\overset{\ominus}{\text{O}}_3\text{H} \\
| \\
\text{C}{=}\text{O} \\
| \\
\text{CH}_2\text{OH}
\end{array}} \\
+ \\
\underset{\text{(G-3-P)}}{\begin{array}{c}
\text{CHO} \\
| \\
\text{CHOH} \\
| \\
\text{CH}_2\text{OP}\overset{\ominus}{\text{O}}_3\text{H}
\end{array}}
\end{array}
\qquad (11\text{-}1)
\end{array}
$$

$$
\begin{array}{ccc}
& \overset{\ominus}{\text{CH}_2\text{OPO}_3\text{H}} & \\
\text{CH}_2\overset{\ominus}{\text{OPO}_3\text{H}} & | & \\
| & \text{C}{=}\text{O} & \\
\text{C}{=}\text{O} & | & \\
| & \text{CH}_2\text{OH} & \\
\text{HOCH} \quad \underset{}{\overset{\text{FDP-A}}{\rightleftharpoons}} & \text{(DHAP)} & \\
| & + & \qquad (11\text{-}2) \\
\text{CHOH} & \text{CHO} & \\
| & | & \\
\text{CHOH} & \text{CHOH} & \\
| & | & \\
\text{CH}_2\text{OH} & \text{CH}_2\text{OH} & \\
\text{(F-1-P)} & \text{(G)} &
\end{array}
$$

Both reactions produce dihydroxyacetone phosphate (DHAP); Eq. 11-1 also produces D-glyceraldehyde-3-phosphate (G-3-P), while Eq. 11-2 produces D-glyceraldehyde (G) in addition to the DHAP. In this chapter the pyranose and furanose ring forms of the sugars will not be used since it seems certain that the open chain form is utilized by the enzyme.

As mentioned, there are two mechanistic types of FDP-aldolases, commonly referred to as the Class I and Class II aldolases. It appears that all plant and animal FDP-aldolases are of the Class I type; most, if not all microbial FDP-aldolases are of the Class II type. Class I enzymes are the Schiff base-forming aldolases and Class II enzymes are the metalloaldolases, that is, all plant and animal FDP-aldolases operate through Schiff base formation between the ketone of FDP and the ε-amino group of a lysine residue in the protein, and they do not require an essential heavy metal atom; all microbial aldolases do not involve Schiff base formation with the substrate and do contain an essential heavy metal, usually Zn^{2+}. This chapter will discuss the rabbit muscle enzyme as an example of the Class I aldolases and the yeast enzyme as an example of the Class II aldolases. Before considering their differences, however, their many points of mechanistic similarity will be discussed.

The General Mechanism

It is well accepted that aldol condensation reactions and their reverse reactions (the retrograde aldol reaction) generally proceed as shown in Eq. 11-3. This reaction has been well studied, and its mechanism is thoroughly treated in beginning organic chemistry courses. The reaction conditions often involve moderately to strongly basic conditions for periods of time over an hour and occasionally at elevated temperatures in order to maximize yields. To carry out the retrograde reaction for glycolysis, nature had to evolve

$$R-CH_2\overset{|}{\underset{R''}{C}}=O + \ddot{B} \rightleftharpoons$$

$$R\overset{\ominus}{C}H-\overset{|}{\underset{R''}{C}}=O + BH^{\oplus} \xrightleftharpoons{R'CH_2CHO} R'CH_2-\overset{O^{\ominus}}{\underset{H}{\overset{|}{C}}}-\overset{|}{\underset{R}{CH}}-\overset{|}{\underset{R''}{C}}=O$$

$$R'CH_2\overset{OH}{\underset{H}{\overset{|}{C}}}-\overset{|}{\underset{R}{CH}}-\overset{|}{\underset{R''}{C}}=O \underset{-H^{\oplus}}{\overset{+H^{\oplus}}{\rightleftharpoons}} \quad (11\text{-}3)$$

a means of stereospecifically abstracting a proton to form the alkoxide anion, provide a means to ensure carbanion formation, and do all this under normal physiological conditions. In plants and mammals the solution has been the Schiff base forming aldolases, in microorganisms, the metallo-aldolases.

Although the detailed mechanisms of glycolysis are different for the two classes, experimental observations on both classes of aldolases point to the utilization by the enzymes of essentially the same overall mechanism as the standard organic reaction, redrawn in Eq. 11-4 for the reversible cleavage of

$$\begin{array}{c} CH_2OPO_3^{-2} \\ | \\ C=O \\ | \\ HOCH \\ | \\ CHOH \\ | \\ CHOH \\ | \\ CH_2OPO_3^{-2} \\ FDP \end{array} + \begin{bmatrix} -B_2H^{\oplus} \\ -B_1 \end{bmatrix} \rightleftharpoons \begin{bmatrix} CH_2OPO_3^{-2} \\ | \\ C=O\cdots\begin{bmatrix} -B_2H^{\oplus} \\ -B_1H^{\oplus} \end{bmatrix} \\ | \\ HOCH \\ | \\ CHO^{\ominus} \\ | \\ CHOH \\ | \\ CH_2OPO_3^{-2} \end{bmatrix} \rightleftharpoons \begin{bmatrix} CH_2OPO_3^{-2} \\ | \\ C=O\cdots\begin{bmatrix} -B_2H^{\oplus} \\ -B_1H^{\oplus} \end{bmatrix} \\ | \\ HOCH \\ \ominus \end{bmatrix}$$

$$\begin{array}{c} \mathbf{I} \\ + HC=O \\ | \\ CHOH \\ | \\ CH_2OPO_3^{-2} \\ G\text{-}3\text{-}P \end{array}$$

$$\begin{array}{c} CH_2OPO_3^{-2} \\ | \\ C=O \\ | \\ CH_2OH \\ DHAP \end{array} + \begin{bmatrix} -B_2H^{\oplus} \\ -B_1 \end{bmatrix} \rightleftharpoons \begin{bmatrix} CH_2OPO_3^{-2} \\ | \\ C=O\cdots\begin{bmatrix} -B_2 \\ -B_1H^{\oplus} \end{bmatrix} \\ | \\ CH_2OH \end{bmatrix} \quad (11\text{-}4)$$

FDP.[2] For example, in both enzymes one can observe exchange of H atoms at C3 of DHAP with the solvent water. This is studied by using deuterated or tritiated water, incubating the enzyme with DHAP, isolating the DHAP after a suitable time, and examining it for incorporated D or T. The observation of exchange implies the intermediacy of the carbanion I in Eq. 11-4. For both enzymes, the exchange of T into DHAP at C3 is inhibited by the presence of FDP or G-3-P, suggesting, but not proving, that the same site on the enzyme is utilized for the cleavage and synthesis reactions and the exchange reaction. To prove the identity of the site minimally requires a determination of whether the inhibition is competitive or not. If the sites are identical, competitive inhibition should be observed (see Chapter 6). Only one H atom of DHAP at C3 is exchangeable with D or T, and when FDP is synthesized from unlabeled DHAP and G-3-P in D_2O, no deuterium is incorporated into the product FDP. These two facts demonstrate that the exchange reaction is stereospecific and that by virtue of its mode of binding of DHAP, the enzyme can distinguish between the two H atoms present at C3.[3]

As we will see later, the most plausible explanation for the role of Schiff base formation in the Class I aldolases and of the Zn^{2+} in the Class II aldolases is that in either case an electron sink is provided to stablize the carbanion at C3, allowing the reaction to proceed at neutrality and 37°C.

Through the use of C^{14} labels on both DHAP and G-3-P it was shown that these trioses bind independently to FDP-A. This was done by incubating the enzyme with either C^{14}-DHAP + cold G-3-P or cold DHAP + C^{14}-G-3-P and then measuring the rate of appearance of labeled FDP. If sequential (ordered) binding were required, then both C^{14} incorporation rates would be the same. The different rates that were observed means that both G-3-P and DHAP bind independently, and when both are present synthesis can occur.

Both enzymes are specific for DHAP but somewhat nonspecific as to the condensing aldehyde; both enzymes stereospecifically produce FDP and not its C3 enantiomer.

Class I Aldolases (Schiff Base Aldolases)

Rabbit skeletal muscle aldolase is perhaps the most widely studied of the Class I aldolases that have been isolated. Its isolation from the tissue is quite simple.[4] The muscle is passed through a meat grinder, the ground tissue is extracted twice with an equal weight of water, and the pH is adjusted to 7.5 with NaOH. An equal volume of saturated $(NH_4)_2SO_4$ is added, followed by adjustment of the pH back to 7.5 with NH_4OH. This is cooled to 0–5°C, and the precipitate is filtered. The supernatant is brought to 0.52 saturation in $(NH_4)_2SO_4$; crystals form on standing. When crystals appear, the solution is placed in the cold and allowed to sit for three days. The crystallized enzyme is then collected by filtration and recrystallized from 0.52 saturated $(NH_4)_2SO_4$ in a total volume such that the protein concentration is kept below 2%.

The protein has a molecular weight of 160,000 and the catalytic unit consists of four subunits of molecular weight 40,000. This point has recently received some close attention since the muscle enzyme is partially dissociated at physiological concentrations (0.02 mg/ml maximum). When ultracentrifugal studies were done in $1.5M$ urea where the enzyme is known to be dissociated but active, it was found that in the absence of FDP substrate, s_{20w} was depressed; but when FDP was present, s_{20w} for the ES complex was 7.2. This is sufficiently close to the value for the tetramer (7.4 to 7.8 depending on conditions) that it seems clear that the tetramer is the active enzyme.[5]

The single most characteristic mechanistic property of the Class I aldolases is their formation of a Schiff base between the ketone carbon of FDP or DHAP and the ε-amino group of a lysine residue. In fact, each subunit will do this, for chemical modification studies have shown each subunit to contain an active site. Schiff base formation was recognized when it was found that $NaBH_4$ inactivated the enzyme completely if incubated with the enzyme in the presence of DHAP.[2] Degradation of the protein and isolation of ε-N-β(1,3-dihydroxypropyl)lysine (**II**) provided very strong evidence for the presence of the Schiff base intermediate **III** on the enzyme prior to the

reduction. Moreover, the unusually facile $NaBH_4$ reduction suggests that the enzymic Schiff base may well be protonated within the active site. This would imply an abnormal environment for that protonated Schiff base since its pK_a would normally be expected to be between pH 5 and 7,[6] but the rabbit muscle aldolase pH profile shows a broad optimum from pH 6.9 to about 9. If the Schiff base is protonated, it might be expected to more effectively stabilize the carbanion intermediate **I** in Eq. 11–4. (See also the discussion on the Class II enzymes.)

As with many other enzymes, the active site is considered to be apolar and essentially nonaqueous.[2] Functionally, this helps prevent protonation of the carbanionlike intermediate, which would occur rapidly in the presence of water.

A variety of other chemical modification studies (besides borohydride reduction) have been performed on rabbit muscle FDP-A and in one way or another have implicated cysteine, histidine, and tyrosine in the enzymic process.[2] For example, bromopyruvic acid was found to alkylate six sulfhydryl groups and cause 100% inactivation of the enzyme. However, in the presence of the substrate FDP, no loss of activity was observed and only two to three sulfhydryl groups were alkylated.[7] This suggests that one SH group per subunit is functionally involved in the catalytic mechanism.

Photosensitized oxidation of histidine residues accomplished the destruction of between 18 and 28 histidine residues (50–70%) and a loss of between 80 and 90% of the initial activity.[8] Schiff base formation with DHAP and FDP still occurred, but the rate of exchange of T into C3 of DHAP was significantly lowered, so that the protonation/deprotonation step at C3 becomes rate determining. Normally for aldol condensations carbon-carbon bond formation (or rupture) is rate determining. Both FDP and F-1-P cleavage rates are reduced by the photooxidation. On the other hand, removal of three out of four C-terminal tyrosine residues by treatment with carboxypeptidase A (CP-A) caused a dramatic loss of FDP cleavage capacity but none for F-1-P.[9] In the reversible FDP cleavage reaction, it was found that the CP-A treated enzyme exchanged C^{14}-G-3-P into FDP at a normal rate but C^{14}-DHAP was incorporated at a much reduced rate, implying that the result of the CP-A treatment was also to make protonation/deprotonation at C3 rate determining.

On the basis of these and other experiments Scheme 11-1 can be proposed as an incomplete mechanism for rabbit muscle-FDP-aldolase action. It is clear from the tyrosine modification studies that at least two basic groups at each active site are involved—possibly one for each of the two proton transfers. If only one group was involved in both proton transfers, then the CP-A treatment would have affected both C^{14}-DHAP *and* C^{14}-G-3-P incorporation into FDP, but only the former was decreased by the treatment. The portion of the reaction involving DHAP requires carbanion formation at C3 of DHAP, and, as mentioned, the rate of this step was significantly reduced by the CP-A treatment. The portion involving G-3-P requires alkoxide formation at C4 of the newly formed hexose and its subsequent protonation. The same modification that reduced the rate of deprotonation of DHAP had no effect on the rate of protonation of the alkoxide, and this implies that a different basic group may be involved in this step.

The fact that both tyrosine and histidine modification affected the same step (protonation/deprotonation at C3) implies either that one of the groups controls the conformational integrity of the enzyme while the other is involved in proton donation, or that a "change-relay" system, analogous to that proposed in chymotrypsin studies,[10] involving both the important histidine and tyrosine residues (and other residues), and necessary for the C3 protonation/deprotonation, is disrupted by the modifications.

It is clear that much remains to be done with the Class I aldolases. Perhaps the most important question is the identity of the prototropic groups for each proton transfer. A related question is the functional importance of the groups that have been implicated in the mechanism from the chemical modification studies discussed earlier. In addition, the existence of the protonated Schiff

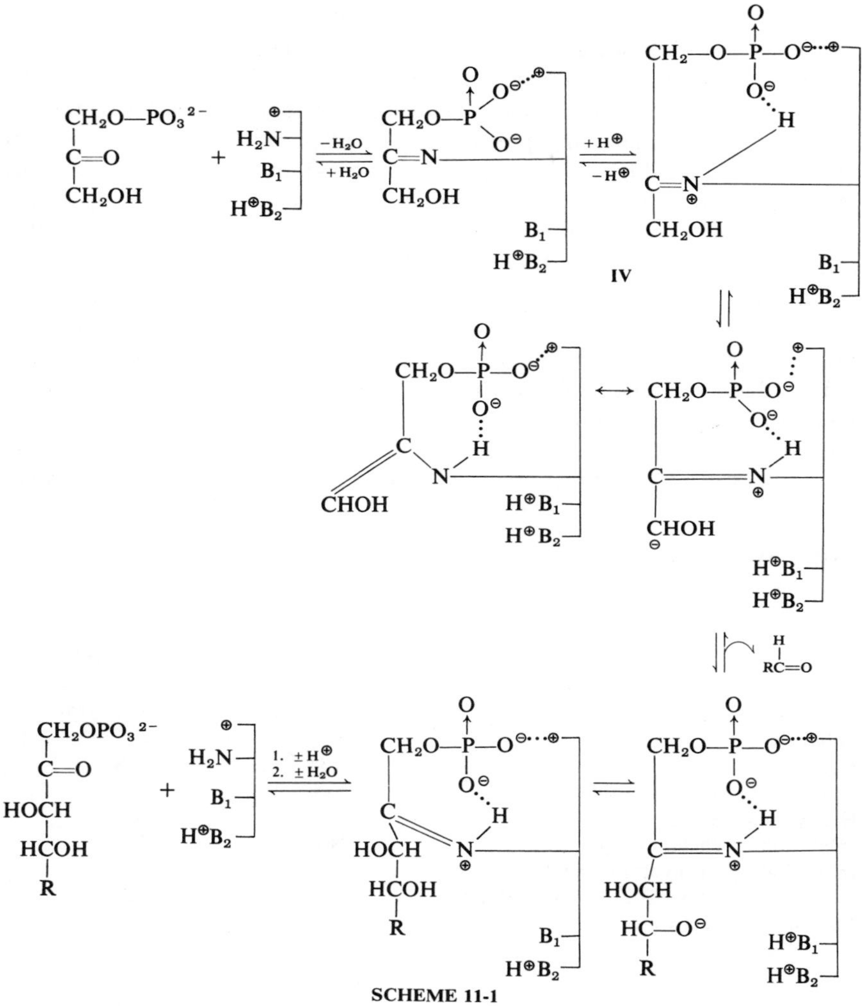

SCHEME 11-1

base intermediate **IV** is still conjectural. As shown, its presence would allow resonance stabilization of the carbanion intermediate—which is shown as a mechanistic representation of the electron sink function of the Schiff base.

Class II Aldolases (Metalloaldolases)

Among the metalloaldolases, the enzyme isolated from yeast is one of the more widely studied enzymes; hence we will discuss its properties as an example of this class. As with the rabbit muscle enzyme, the isolation of the yeast enzyme is quite simple.[11] The yeast is lysed by stirring in warm toluene 1–1.5 hours followed by standing at room temperature in toluene for an additional 2 hours. The toluene autolysate is extracted with water to transfer the enzyme to the aqueous phase. This aqueous phase is then fractionated with $(NH_4)_2SO_4$, removing precipitates at 50% and 70% saturation. The enzyme is obtained by then bringing the supernatant to 80% saturation with $(NH_4)_2SO_4$ and allowing the solution to stand in the cold overnight. The precipitate now obtained by centrifugation is crystallized first from 25% $(NH_4)_2SO_4$ and then from 10–15% $(NH_4)_2SO_4$.

During this process it is very advantageous to keep all aqueous solutions or suspensions of the enzyme $0.05 M$ in β-mercaptoethanol. This prevents inactivation that might be caused by oxidation of essential sulfhydryl groups.

In Table 11-1, many of the characteristics of both the yeast and rabbit muscle aldolases are listed for comparison.[2,11–13] An important difference immediately noted is that the yeast enzyme is completely insensitive to $NaBH_4$. This makes it highly unlikely that a Schiff base is formed during the catalytic

TABLE 11-1 Properties of Class I and Class II Aldolases[a]

Property	Class I (Rabbit Muscle Aldolase)	Class II Yeast Aldolase
Molecular weight	160,000	70–80,000
Subunits	4	2
s_{20w}	7.8S	5.4S
Inactivation by borohydride	Yes	No
Heavy metal requirement	None	Zn^{2+}
Inhibition by chelating agents	No	Yes
Inhibition by sulfhydryl reagents	Yes	Yes
Partial activity loss from CP-A treatment	Yes	No
Potassium ion activation	No	35-fold (2- to 10-fold for other examples)
pH Optimum:		
FDP Cleavage	6.9–8.8	6.9–7.1
T exchange into DHAP	6.9–8.8	6.0
$K_m(M)$		
FDP	6×10^{-5}	3×10^{-4}
DHAP	2×10^{-3}	2×10^{-3}
G-3-P	1×10^{-3}	1×10^{-3}
V_{max} (FDP cleavage) M/liter/min/M enzyme	5300	12,500
FDP/F-1-P Specificity	50	About 10^4

[a] Data from references 2, 11, 12, and 13.

process of the Class II enzymes. The EPR (electron paramagnetic resonance) studies discussed later show that the metal ion of the metalloaldolases is intimately associated with the carbonyl oxygen of DHAP (and presumably FDP). When this fact is combined with the borohydride insensitivity, it becomes highly unlikely that Schiff base formation occurs. The *remote* possibility still exists that a Schiff base formed is inaccessible to the borohydride for steric reasons.

The yeast enzyme has a molecular weight of approximately 70,000 and consists of two apparently equivalent subunits. It has an absolute requirement for 1 Zn^{2+}/mole (metalloaldolases from other species occasionally utilize a different metal ion). The removal of the metal ion by chelating agents such as EDTA leads to a 100% loss of activity. This loss is reversible and activity is restored upon readdition of the necessary amount of Zn^{2+}. This fact has been used by Mildvan et al.[14] to allow them to attempt to regenerate activity by replacing the Zn^{2+} with Mn^{2+}. The reason for doing this is that Mn^{2+} is paramagnetic and therefore would allow detailed EPR studies to be carried out. Activity can be restored in this way, and the EPR studies on the manganese enzyme have been quite revealing. From measurements of EPR relaxation times, distance calculations have been made that show that the protons of C1 and C3 in both acetol phosphate and DHAP are equidistant from the Mn^{2+} atom in the Mn-FDP-A. This is interpreted to mean phosphoryl *and* carbonyl coordination to Mn^{2+} (and presumably to Zn^{2+} in the native enzyme) for normal substrates. The interaction of Mn^{2+} with DHAP or acetol phosphate in the absence of the apoenzyme showed the C1 protons to be significantly closer (40%) to the Mn^{2+} than were the C3 protons, implying that without the protein present only the phosphoryl group coordinated with the metal ion. Figure 11-1 is a schematic representation of this, showing the proposed bidentate liganding arrangement. Structure *c* of Figure 11-1 is an extrapolation of these findings to the Class I enzymes, which seems quite sound since this type of structure would quite likely have a much higher pK_a than the nonhydrogen bonded protonated Schiff base. If indeed this is the structure of the Schiff base, then the pH profile for the Class I aldolases becomes more understandable (see below).

Figures 11-2 and 11-3 show the comparative pH profiles for the two types of enzymes, and profound differences are seen.[13] Not only do the Class II aldolases have much sharper pH-rate profiles for both DHAP-T exchange and FDP cleavage, but the two reactions have different pH optima. For the Class I enzymes, the exchange and cleavage pH-rate profiles coincide. The reason for the different pH optima for the two reactions with the Class II enzymes has not yet been explained.

As suggested from the use of β-mercaptoethanol in the purification procedure, sulfhydryl groups are important for enzymatic function. This has

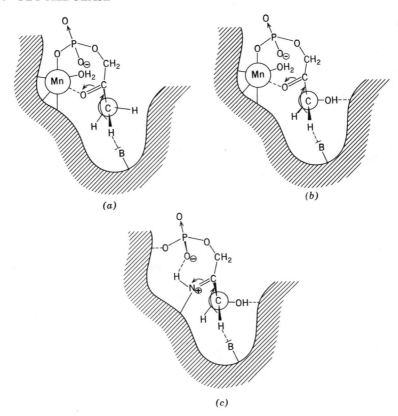

FIGURE 11-1. Probable structures and mechanisms of yeast aldolase-substrate complexes. The manganese bridge complexes of acetol phosphate (*a*) and DHAP (*b*) with yeast aldolase, respectively. A proposed mechanism (*c*) for the Class I aldolases by analogy with structures (*a*) and (*b*). In all cases, B represents a basic group on the enzyme.

been demonstrated by the inhibition of the enzyme with sulfhydryl reagents.[2]

Another distinctive feature of the metalloaldolases is the marked stimulation of activity by potassium ions. This is particularly dramatic for the yeast enzyme where the enhancement factor is 35-fold. For many other metalloaldolases, this factor is between threefold and tenfold. This phenomenon is not at all understood.

On the basis of the modification studies, the EPR studies and exchange and kinetic studies, the mechanism in Scheme 11-2 may be proposed. In this scheme, the structures **V** are a representation of the electron sink function of the zinc ion. Since this enzyme has not been studied in as great detail as the muscle enzyme, one cannot be as explicit about the number of basic

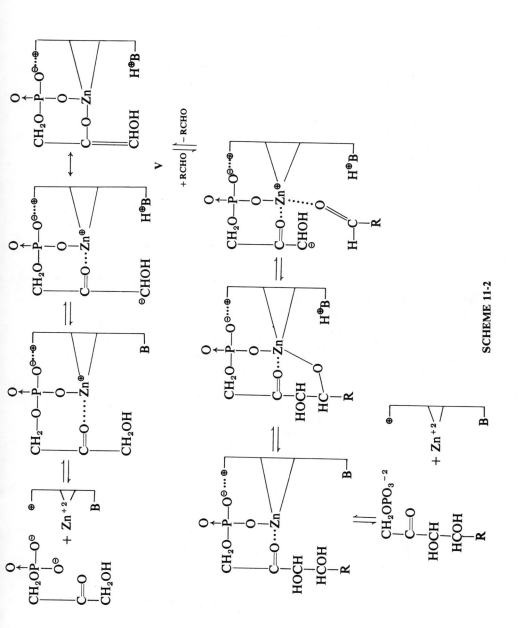

SCHEME 11-2

251

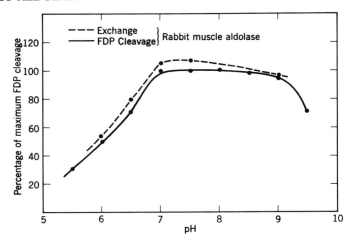

FIGURE 11-2. pH profile for Class I aldolase activity.

groups involved in the reaction. Hence only one basic group is shown as being involved in both proton transfers. The pK_a values for prototropic groups that may be estimated from the pH-rate profiles are about 6.5 and about 8 for FDP cleavage and about 5.8 and 6.8 for exchange. However, it is unclear whether these represent four, three, or two different prototropic species, since the shifts could be due to unusual microenvironmental effects occurring during binding and catalysis. Moreover, even though the ionizations

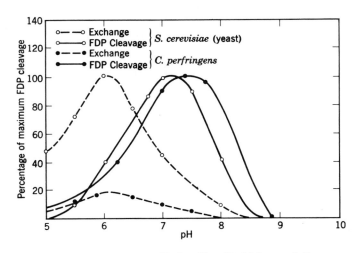

FIGURE 11-3. pH profile for Class II aldolase activity.

of two groups are the minimum requirement for a bell-shaped pH-rate profile (see Chapter 5), the results with chymotrypsin (Chapter 9) imply that one or both of these may govern conformational changes crucial to activity and not control a step in the catalytic process.

Thus it is obvious that on this enzyme as well as the Class I aldolases, much more work remains before the mechanism is completely understood. For example, in addition to the unanswered questions pointed out throughout this chapter, very little is known about the subunit interactions of these enzymes—particularly those of Class II—and their relation to catalysis. Since the enzymes of glycolysis and glycogenesis are extremely important enzymes, it is to be expected that studies aimed at increasing our understanding of aldolases will continue.

References

1. Report of the Commission on Enzymes, I.U.B. Symposium Series, Vol. 20, Pergamon Press, New York, 1961, p. 30.
2. D. E. Morse and B. L. Horecker, *Adv. Enzymol.*, **31**, 125 (1968).
3. W. J. Rutter and K. H. Ling, *Biochem. Biophys. Acta.*, **30**, 71 (1958).
4. J. F. Taylor, *Meth. Enzymol.* **1**, 319 (1955).
5. C. J. Masters and D. J. Winzor, *Biochem. J.*, **121**, 735 (1971).
6. W. P. Jencks, *Catalysis in Chemistry and Enzymology*, McGraw-Hill, New York, 1969 p. 73.
7. C. Y. Lai and P. Hoffee, *Fed. Proc.*, **25**, 408 (1966).
8. P. Hoffee, C. Y. Lai, E. L. Pugh, and B. L. Horecker, *Proc. Nat. Acad. Sci. U.S.*, **57**, 107 (1967).
9. E. R. Dreschler, P. D. Boyer, and A. G. Kowalsky, *J. Biol. Chem.*, **234**, 2627 (1959).
10. D. M. Blow, J. J. Birktoft, and B. S. Hartley, *Nature*, **221**, 337 (1969).
11. W. J. Rutter, J. R. Hunsley, W. E. Groves, J. Calder, T. V. Rajkumar, and B. M. Woodfin, *Meth. Enzymol.*, **9**, 479 (1966).
12. W. J. Rutter, in *The Enzymes*, Vol. 5, P. D. Boyer, H. Lardy, and K. Myrback (Eds.), Academic, New York, 1961, p. 341.
13. W. J. Rutter, *Fed. Proc.*, **23**, 1248 (1964).
14. A. S. Mildvan, R. D. Kobes, and W. J. Rutter, *Biochemistry*, **10**, 1191 (1971).

CHAPTER 12

Allosterism

It is one of the foremost fundamental tenets of all chemistry that the structure of the individual molecules of a substance determines the measurable macroscopic properties (both chemical and physical) of that substance. And as a corollary to this basic tenet it is accepted that the characterization of microscopic molecular architecture may follow deductively from appropriate experimental observations of macroscopic properties of substances.

We have already dealt implicitly with this very fundamental concept in our discussions of enzyme mechanisms both in general and in particular. For example, the active-site hypothesis which has been central to our consideration of enzyme function is just a special case of this general structure-function correlation in which we maintain that the catalytic powers of an enzyme may be substantially understood in terms of the structural characteristics of the active site of that enzyme and the structure of molecules which interact with that active site.

In this chapter we examine some more complex situations in which enzyme molecules engage in mutual interactions with each other or with small "effector" molecules in such a way that structural or conformational changes of functional importance are induced in the protein. We will find that such cooperative or "allosteric" interactions play a fascinating role in the self-regulation by which living cells control the processes of chemical synthesis and degradation essential to their efficient operation.

Regulatory Enzymes and Metabolic Control

Our discussion of the phenomenon of allosterism in terms of protein structure and function will be more meaningful to us if we first briefly outline the dramatic role played by regulatory phenomena of various kinds in the control of metabolic processes.[1,2]

A living cell can be regarded as a totally automated chemical factory whose job it is to manufacture certain products which are necessary for the continued operation of that cell and the organism of which it is a part. These products are produced from nutrient substances by means of series of degradative and/or synthesizing reactions (the catabolic and metabolic pathways), which constitute the production lines of the cell. In order to make the most efficient use of the raw materials and energy available to it, the cell has evolved an array of feedback circuits which operate to control both the supply of production line machinery (the enzymes) and the rate at which this machinery operates.

Figure 12-1 is a schematic diagram of a portion of a cell production line (metabolic sequence) whose job it is to produce product E from starting

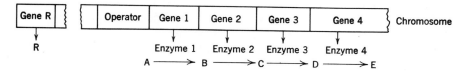

FIGURE 12-1. Schematic diagram of a portion of a cell production line.

material A. Each station along the production line consists of an enzyme which catalyzes the indicated conversion. In addition the production of the necessary enzymes is directed by individual genes found in the chromosomal material (DNA) of the cell nucleus. When some key product E (in this example, the *end product* of the sequence but not necessarily so in all cases) is absent or present in low concentrations, the production line operates at full efficiency. But as the concentration of product E builds up, the whole process begins to slow down until finally, in the presence of enough E, the process may for all practical purposes *stop* altogether. How is it that the presence of product E calls a halt to the further production of E? Two mechanisms are apparently in operation here.

On the one hand, the molecule E may be capable of switching off the genes that direct the synthesis of the required enzymes. Jacques Monod and his colleagues at the Pasteur Institute in Paris discovered this phenomenon, which is called *repression*: the inhibition of enzyme synthesis in the presence of a metabolic product, the product serving as the signal that the enzymes are not needed. According to the Monod interpretation, a given set of "structural" genes in a chromosome (e.g., genes 1–4 in Figure 12-1) is under the control of an adjacent "operator." This operator, a special structure linked closely on the chromosome with the structural genes it controls, is in turn capable of interacting with a special regulatory molecule whose synthesis by the cell is directed by some remote (on the chromosome) "regulatory" gene (gene R in Figure 12-1). Control is ultimately due to the regulatory metabolite (product E in our example) in the following way. The regulatory metabolite E binds to the repressor molecule R to form a complex E·R. This complex E·R (but not R by itself) then binds the genetic operator, thereby switching off the adjacent structural genes. This whole control mechanism operates through binding equilibria according to the simple law of mass action with the concentration of E serving as the determining factor.

It should be noted at this point that the same kind of mechanism has been held accountable for just the opposite phenomenon, enzyme *induction*. That is, cells can also respond to chemical signals calling on them to *produce* enzymes they need. In this case the regulatory metabolite might well be the molecule A rather than E, and instead of activating the repressor R toward binding the operator, A would deactivate it. That is to say, in the case of *induction*, the operator switches off the structural genes in the presence of R itself but *not* in the presence of an A·R complex.

In addition to the operation of the control mechanisms of repression and induction which we have just described,* control is also effected by the interaction of regulatory metabolites (e.g., A or E of Figure 12-1) with certain key enzymes themselves. Thus, referring to our original example in which the buildup of product E resulted in the slowdown and eventual shutdown of the production line which produced it, this product E may serve as a *specific inhibitor* of enzyme 1. Hence we would have negative feedback control by *final* product concentration level on the rate of the *first* reaction of the metabolic sequence leading to that final product. (Also, as in the case for *induction* as opposed to *repression* in the case of enzyme synthesis control, regulatory metabolites may act as specific activators toward regulatory enzymes.) These are the kinds of phenomena we are principally concerned with in this chapter. We will examine the unique properties possessed by the so-called regulatory enzymes, which are subject to positive or negative feedback control of this general type. We will also consider current theories suggesting structural interpretations for these unique properties.

Properties and Regulatory Enzymes

Examination of a large number of different regulatory enzymes has revealed that they possess in common certain structural and kinetic characteristics that distinguish them from other enzymes. The most fundamental of these distinguishing characteristics is the susceptibility of regulatory enzymes to inhibition or activation by metabolites which bear little or no structural

---

* The reader will be aware that the very brief and simplified discussion of the control of *de novo* protein synthesis by regulatory metabolites presented here is by no means a complete one. Further discussion concerning these and other regulatory processes of cell metabolism which will not be dealt with in this book may be found in several review articles.[3-6]

resemblance to the respective substrates for these enzymes. The phenomena of inhibition and activation are not themselves at all uncommon with enzyme catalysis in general. But usually inhibitors or activators for a given enzyme are more or less closely analogous structurally to the substrate(s) for that enzyme and their effects can be readily interpreted in terms of an interaction of these various small molecules at or near the active site of the enzyme in question. The observation for the regulatory enzymes that they may be inhibited or activated not only by the usual structural analogs to the substrate but also by other nonsubstrate-related species occasioned the extremely important suggestion by Monod, Changeux, and Jacob[7] that the modulation of enzyme activity must be achieved not just through the binding of substrates and their analogs at the enzyme *active site* but also indirectly through the binding of small molecules at specific *regulatory sites* (allosteric sites), which are distinct from the active site itself and may indeed be located at some remote position on the protein molecule in question. In presenting their hypotheses about regulatory enzymes, Monod and his co-workers proposed that these small molecules which bind at allosteric sites be called *allosteric effectors* and that enzymes subject to the influence of allosteric effectors be called *allosteric proteins* or *allosteric enzymes*. This terminology has now received wide acceptance and the term "allosteric" has come to be used generally to describe the interaction of any small molecules, including substrates, at binding sites other than the catalytically active site of an enzyme.

At this point we are ready to catalog the more important behavioral symptoms of allosteric enzymes which set them apart from the more common enzymes whose behavior we are now familiar with.

1. Allosteric effectors, which are usually key products of a metabolic sequence influenced by the operation of the enzyme in question, are bound to the enzyme and profoundly influence its catalytic activity. Though they are not related structurally to the substrate, their binding to the enzyme results in either activation or inhibition of the enzyme's activity toward the substrate. This activity change may be attributed to a modification of the affinity of the active site for substrate (i.e., K_s), or to a modification of the rate of which the preformed enzyme-substrate complex yields products (i.e., k_{cat}), or to both. Effects such as these, that is, effects on the nature of the interaction of the enzyme with a ligand of one type (e.g., substrate) resulting from the binding to the enzyme of a ligand of some other type (e.g., allosteric inhibitor), are called *heterotropic effects*.

2. Even in the absence of allosteric effectors as such, the response of initial rate v_0 to increases in initial substrate concentration for many allosteric enzymes is anomalous. Instead of the usual hyperbolic curves (Figure 4-5), plots of v_0 versus $[S]_0$ yield sigmoid saturation curves such as curve II of

Figure 12-2. Such a sigmoid curve implies that at least two substrate molecules interact with a single enzyme molecular unit and that the binding of the first in some way facilitates the subsequent binding of the next. This "cooperativity" which attends multiple binding is frequently observed not only for substrates but also for allosteric effector substances as well (curve III, Figure 12-2), suggesting multiple interacting binding sites on one enzyme

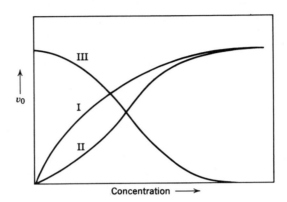

FIGURE 12-2. The influence of substrate and allosteric effector concentration on enzyme activity. Curve I represents normal (Michaelis-Menten) response of activity to substrate concentration. Curve II shows the response of activity to substrate concentration typical of many allosteric enzymes. Curve III shows the response of activity to the concentration of a homotropic allosteric inhibitor (at constant $[S_0]$). (After Stadtman,[1] p. 49.)

molecular unit for a given allosteric effector. Such allosteric effects resulting from the influence of the binding of one ligand (substrate or allosteric effector) on the interaction of the enzyme with other identical ligands are called *homotropic effects*. Homotropic effects are always cooperative (never antagonistic).

3. Agents that might function as simple inhibitors with normal enzymes (i.e., substrate analogs) may, with allosteric enzymes, give rise to quasi-homotropic activation at low concentrations and "normal" competitive inhibition at high concentrations.

4. In many cases, allosteric effects may be correlated with protein quaternary structure. The species referred to previously as the "enzyme molecular unit" may in fact be a polymer or oligomer consisting of several identical monomers (called protomers by Monod and his school), each of which in turn may be made up of subunits which are individual (identical or nonidentical) protein or polypeptide chains. Thus reversible equilibria of the type polymer ⇌ monomers (protomers) ⇌ subunits (where "monomer" represents the minimal structural unit still capable of catalytic activity) must

often be taken into account in interpreting the behavior of allosteric enzymes. For example, various treatments that lead to alterations in protein conformation (and therefore might well be expected to have a profound influence on quaternary structure) may diminish or eliminate allosteric effects (both homotropic and heterotropic) without substantially affecting the *binding* of the individual allosteric effectors or the substrate. Such treatments include[7] treatment of the enzyme with mercuric salts or *p*-chloromercuribenzoate, aging in the cold (0–5°), dialysis, freezing, treatment with urea, treatment with proteolytic enzymes, subjection to high ionic strength, shifts in pH, and heating.

We will now proceed to the consideration of a general model proposed by Monod, Wyman, and Changeux[8] to account for the characteristic features of allosteric enzymes we have just enumerated.

The Model of Monod, Wyman, and Changeux

The model is described by the following statements:[8]

"*1. Allosteric proteins are oligomers the protomers of which are associated in such a way that they all occupy equivalent positions. This implies that the molecule possesses at least one axis of symmetry.*"

"*2. To each ligand able to form a* stereospecific *complex with the protein there corresponds one, and only one, site on each protomer. In other words, the symmetry of each set of stereospecific receptors is the same as the symmetry of the molecule.*"

"*3. The conformation of each protomer is constrained by its association with the other protomers.*"

"*4. Two (at least two) states are reversibly accessible to allosteric oligomers. These states differ by the distribution and/or energy of inter-protomer bonds, and therefore also by the conformational constraints imposed upon the protomers.*"

"*5. As a result, the affinity of one (or several) of the stereospecific sites towards the corresponding ligand is altered when a transition occurs from one to the other state.*"

"*6. When the protein goes from one state to another state, its molecular symmetry (including the symmetry of the conformational constraints imposed upon each protomer) is conserved.*"

Let us assume that we have an allosteric protein made up of four identical (as required by Statement 1) protomers, each of which has three stereospecific binding sites, one for substrate S, one for inhibitor I, and one for activator A (Statement 2). The *free* protein is present in two states, which we shall call R_0 and T_0, and which are in equilibrium (Statement 4). An R_0 to

T_0 conversion (or vice versa) will result in a conformational change in all four protomers, and all four will change in exactly the same way (Statements 3 and 6). Because of this conformational change, the affinity of a given set of four equivalent binding sites for the corresponding allosteric ligand (S, I, or A) will differ for the two different states (Statement 5).

Cooperative homotropic effects can now be accounted for as follows. Consider the substrate S. Let us assume that it binds to the substrate site in R state protomers with a dissociation constant K_R and to T state protomers with a dissociation constant K_T, and that due to the conformational differences between R and T states, $K_R/K_T \ll 1$. That is, the R state is much more attractive to S than the T state is. Let us also assume that the equilibrium constant L for the $R_0 \rightleftharpoons T_0$ interconversion is large. That is, in the absence of any ligand molecules, only a small fraction of the total available allosteric protein is in the more active form R_0, which has the greater affinity for substrate.

Now what does the model predict will happen when the ligand species S is added? The following successive equilibria are established:

$$R_0 + S \xrightleftharpoons{K_R^{-1}} RS_1 \qquad R_0 \xrightleftharpoons{L} T_0 \qquad T_0 + S \xrightleftharpoons{K_T^{-1}} TS_1$$

$$RS_1 + S \xrightleftharpoons{K_R^{-1}} RS_2 \qquad\qquad\qquad TS_1 + S \xrightleftharpoons{K_T^{-1}} TS_2$$

$$RS_2 + S \xrightleftharpoons{K_R^{-1}} RS_3 \qquad\qquad\qquad TS_2 + S \xrightleftharpoons{K_T^{-1}} TS_3 \qquad (12\text{-}1)$$

$$RS_3 + S \xrightleftharpoons{K_R^{-1}} RS_4 \qquad\qquad\qquad TS_3 + S \xrightleftharpoons{K_T^{-1}} TS_4$$

Given the relative magnitude of K_R and K_T, added S is most likely to bind R-state sites, even though there are far more T-state sites available initially. Whenever an S molecule binds to an R_0 protein, two very significant conditions result: (1) as long as one or more S molecules remain bound, the protein molecule remains "frozen" in the R state; (2) the $R_0 \rightleftharpoons T_0$ equilibrium is perturbed by the removal of an R_0 molecule at the expense of T_0 (LeChatelier's principle). These two conditions imply cooperativity (homotropic allosteric effects) because the result of binding S to R_0 is to *produce a net increase in the total number of R-state binding sites for S*, and hence a net increase in the affinity of the total protein for S.* At low substrate concen-

---

* This is perhaps a difficult point to visualize. Let us consider a hypothetical case as an example of what is implied here. Take an allosteric protein where the allosteric constant L is such that for every R_0 molecule there are 99 T_0 molecules. That is to say, in the absence of any added substrate or other allosteric effector, only one out of 100 enzyme molecules is in the "active" form. If our enzyme is a tetramer, this means that out of 400 *potential* active sites, only four are really "active." Now if we add a substrate molecule and it binds to one of these four sites on an R_0 molecule, the $R_0 \rightleftharpoons T_0$ equilibrium tends to readjust to provide a *new* R_0 molecule. Thus the second added substrate molecule sees

trations, the cooperativity is *strong* because R_0 is the only species present in significant amounts to which S will bind favorably. As [S] increases, cooperativity tends to level off and then decrease as RS_1, RS_2, and RS_3 states increasingly compete with a decreasing supply of R_0 for unbound S molecules.

The predictions of this model can be expressed more precisely in mathematical terms. Monod, Wyman, and Changeux[8] define a *saturation function* $\overline{Y}_F$ for any given allosteric ligand F (F = S, I, or A) as the fraction of sites actually occupied by the ligand,

$$\overline{Y}_F = \frac{(RF_1 + 2RF_2 + \cdots + nRF_n) + (TF_1 + 2TF_2 + \cdots + nTF_n)}{n[(R_0 + RF_1 + RF_2 + \cdots + RF_n) + (T_0 + TF_1 + TF_2 + \cdots + nTF_n)]}$$

where n = number of protomers per allosteric protein molecule. If we define α and c such that $F/K_R = \alpha$ and $K_R/K_T = c$ we can express $\overline{Y}_F$ as a function of the variable α:

$$\overline{Y}_F = \frac{Lc\alpha(1 + c\alpha)^{n-1} + \alpha(1 + \alpha)^{n-1}}{L(1 + c\alpha)^n + (1 + \alpha)^n} \tag{12-3}$$

Figure 12-3 shows how the saturation function $\overline{Y}_F$ varies with ligand concentration ($\alpha \propto F$) under various assumed circumstances. Note that cooperativity is greatest when L is large and c is small. Note also that if $c = 1$ (i.e., if $K_R = K_T$ meaning the R and T states have equal affinity for ligand F) or if $L \to 0$ (i.e., if virtually all the free enzyme is in the state which favors F binding), then the saturation function simplifies to

$$\overline{Y}_F = \frac{\alpha}{1 + \alpha} = \frac{F}{K_R + F} \tag{12-4}$$

This is just the Michaelis-Menten equation where $\overline{Y}_F = v_0/V_{max} = v_0(k_{cat} \cdot E_0)$ and $K_R = K_m$.

To this point we have considered the predictions of the Monod-Wyman-Changeux model vis-à-vis only homotropic cooperative interactions of allosteric ligands. The model also accounts for heterotropic effects. In the simplest case, we might have an allosteric protein where substrate S and

---

essentially just as many R_0 molecules as the first did, *but in addition it has available to it the three remaining unoccupied active sites of the RS_1-state enzyme to which the first substrate molecule is already bound.* As more and more substrate molecules are added, to the extent they bind to R_0 rather than RS_1 enzyme molecules, the effect will be to steadily *increase* the number of active (R-state) substrate binding sites. Eventually a point is reached where the number of available substrate-binding sites in RS_1, RS_2, . . ., RS_{n-1} enzymes exceeds those in R_0 enzymes. At this point, cooperativity (binding of substrate increasing affinity of remaining protein for more substrate) falls off.

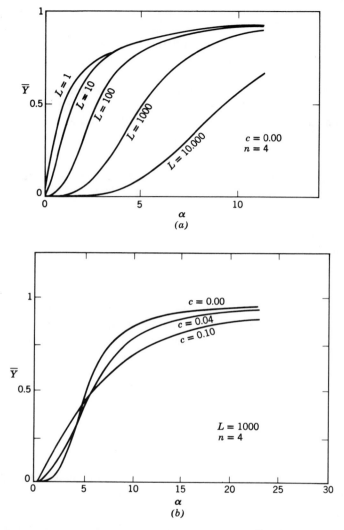

FIGURE 12-3. Theoretical curves of the saturation function $\bar{Y}$ drawn to various values of the constants L and c, with $n = 4$ (i.e., for a tetramer).

activator A have significant affinity for their respective binding sites only when the protein is in the R state. On the other hand, inhibitor I binds significantly only to the T state. Consider the effect of the presence of inhibitor I on the saturation of enzyme with S. Since I binds only to the T state, adding I has the effect of displacing the $R_0 \rightleftharpoons T_0$ equilibrium to the right as T_0 molecules are bound by I to form TI_1, TI_2, . . ., TI_n complexes. This of course reduces the availability of R_0 protein molecules, which obviously means that substrate molecules are faced with a shortened supply of R-state protein to which it can bind, hence the inhibition. Adding activator has just the opposite effect. That is, the added activator binds only R-state protein and consequently displaces the $R_0 \rightleftharpoons T_0$ equilibrium to the left. Thus in the presence of activator substrate sees more R-state protein (has a greater than normal fraction of the total S binding sites available to it).

Again, the situation just described can be analyzed and expressed mathematically. This time, the saturation function for substrate in the presence of activator and inhibitor takes the form

$$\overline{Y}_s = \frac{\alpha(1 + \alpha)^{n-1}}{L(1 + \beta)^n/(1 + \gamma)^n + (1 + \alpha)^n} = \frac{\alpha(1 + \alpha)^{n-1}}{L' + (1 + a)^n} \qquad (12\text{-}5)$$

where L and n have the same significance as previously and where

$$\alpha = \frac{S}{K_S}, \qquad \beta = \frac{I}{K_I}, \qquad \text{and} \qquad \gamma = \frac{A}{K_A}$$

Figure 12-4 illustrates the predictions of the model concerning these heterotropic interactions. It is particularly interesting to note that the inhibitor *increases the cooperativity* of the substrate saturation curve. This is as it should be since the effect of I is to increase L', the "apparent" $R_0 \rightleftharpoons T_0$

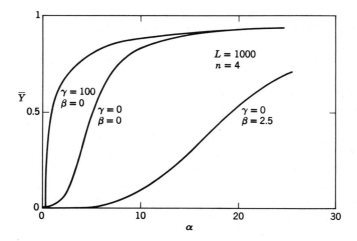

FIGURE 12-4. **Theoretical curves showing the heterotropic effects of an allosteric activator (γ) or inhibitor (β) upon the shape of the saturation function for substrate (α) according to Eq. 12-5.**

equilibrium constant. As mentioned earlier, increases in L imply increases in cooperativity. Of course the activator *decreases* the cooperativity of the substrate saturation curve in effect due to a decrease in L'.

We have seen how the Monod-Wyman-Changeux model accounts at least qualitatively for the kinds of behavior observed with regulatory enzymes as described in the previous section of this chapter, that is, heterotropic and homotropic effects of allosteric ligands. The role of quaternary structure is of course implicit in the very nature of this model; this feature more than anything else accounts for the wide acceptance of this model over others that can account for the kinetic phenomena but do not assign a central role to protein quaternary structure (e.g., see ref. 9). It is by no means yet established that the Monod-Wyman-Changeux model (or some variation of it[10]) is *correct*, but it is in accord with all the currently known facts concerning the structure and function of regulatory enzymes.

Aspartate Transcarbamylase (EC 2.1.3.2)

Although quite a number of regulatory enzymes have been identified and studied,[1] the enzyme aspartate transcarbamylase (ATCase) from *E. coli* was among the first shown to be susceptible to feedback inhibition by a metabolic end product,[11] and it has been most thoroughly investigated. Indeed, the properties of this enzyme more than any other are widely regarded as prototypical of regulatory or allosteric enzymes in general, much as α-chymotrypsin has served as the prototype among proteolytic enzymes. Thus it is appropriate that we conclude this chapter with a consideration of the more significant properties of ATCase in light of the general model for allosteric enzymes we have developed.

ATCase is the first enzyme of a metabolic sequence responsible for the biosynthesis of the pyrimidine nucleotides which are eventually incorporated along with the purine nucleotides into DNA and RNA (Scheme 12-1). The starting materials for this "production line" are carbamyl phosphate and aspartate (which are also, of course, the cosubstrates of ATCase) and the finished products are uridine triphosphate (UTP) and cytidine triphosphate (CTP) (for RNA biosynthesis) and deoxythymidine triphosphate (dTTP) and deoxycytidine triphosphate (dCTP) (for DNA biosynthesis). These pyrimidine nucleotides *are all inhibitors of ATCase*. Here then is an actual example of negative feedback control on a regulatory enzyme by end products. In addition, adenosine triphosphate (ATP) and deoxyadenosine triphosphate (dATP) (purine nucleotides) are specific *activators* for ATCase.[12] Hence we have just the control circuits needed by the cell to regulate production of pyrimidine nucleotides for RNA and DNA biosynthesis. An oversupply of pyrimidine nucleotides inhibits the machinery which produces them.

An oversupply of purine nucleotides, on the other hand, stimulates that machinery to produce at a faster than normal rate to correct a purine-pyrimidine imbalance.

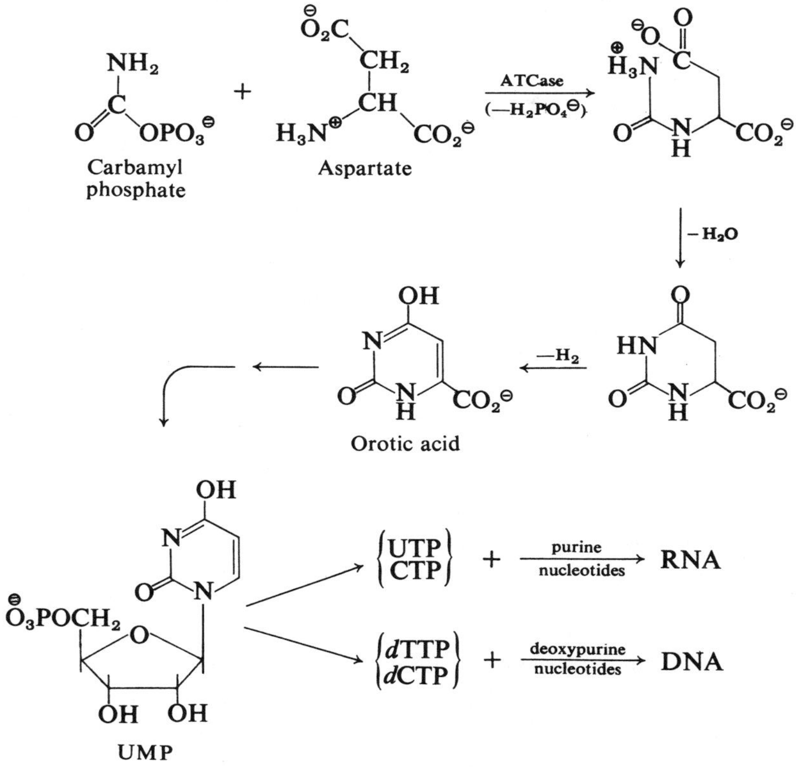

Carbamyl phosphate

Aspartate

Orotic acid

UMP

SCHEME 12-1

Let us see in what observable respects the behavior and properties of ATCase conform to the model we have described. First, the saturation curve (v_0 versus $[S]_0$) for aspartate in the presence of carbamyl phosphate is indeed sigmoidal, indicating homotropic cooperativity.[12] It has also been found that succinate, a structural analog of aspartate though not a substrate, also binds to ATCase in the presence of carbamyl phosphate with a sigmoid saturation curve.[13] Equilibrium dialysis experiments[13] indicate that there are four succinate (or aspartate) binding sites per molecule of the native enzyme.

As mentioned earlier, CTP is a specific inhibitor for ATCase, whose binding to the enzyme is described by a straightforward *noncooperative* (hyperbolic) saturation curve. But there are also four CTP binding sites per molecule of native ATCase.[13] This is also true for the closely similar inhibitor species 5-bromocytidine triphosphate (BrCTP). It is apparent from equilibrium dialysis studies in the presence of BrCTP and succinate together that these two species do *not* simply compete for the same binding sites on the ATCase molecule but bind to their own distinct sites. Yet the binding of succinate and BrCTP, although not simply competitive, is *mutually antagonistic* (heterotropic effect).

Activators for ATCase, such as ATP, *reverse* the inhibition of the enzyme caused by CTP and in fact increase the apparent affinity of ATCase for aspartate even in the absence of any specific inhibitor.[12,13] Equilibrium dialysis experiments show that there is direct *competitive* antagonism between ATP and CTP for presumably *identical* binding sites on native ATCase.[13]

To summarize what we have said so far, it appears that ATCase has four binding sites for substrate and four other binding sites for inhibitor or activator. Homotropic effects are observed with the binding of substrate, but not with the binding of inhibitor or activator. Heterotropic effects are observed with any pair of the three types of ligand.

The most interesting part of the ATCase story has to do with what has been demonstrated concerning the role of subunits and quaternary structure in the behavior we have been discussing. When native ATCase is subjected to certain treatments[12,14] (brief heating to 60°, partial urea denaturation, treatment with mercurials) it dissociates into a collection of two kinds of subunits. Each native enzyme molecule (mol. wt. 310,000) gives rise to four identical subunits of type I (mol. wt. 30,000) and two identical subunits of type II (mol. wt. 96,000). These can be separated and studied independently. The type I subunits (regulatory subunits) are catalytically inactive toward aspartate and carbamyl phosphate, neither of which bind to them. But they *do* bind inhibitors like CTP or activators like ATP on a one-for-one basis. The type II subunits (catalytic subunits) are indeed active toward the substrate pair. In fact they are slightly *more* active than the native enzyme itself, but though each subunit has two substrate binding sites,[13] binding of substrate to these subunits is *not* cooperative. Nor is the activity of the catalytic subunits affected by the presence of CTP or ATP.

In addition to these findings concerning the *independent* subunits of ATCase, experiments have been carried out which clearly indicate that the intact native enzyme is capable of existing in at least two distinct conformational states, a swollen conformation which favors substrate binding and is thus most "active," and a more compact conformation which favors inhibitor binding and is not as "active."[15] These two states of course correspond nicely with the hypothetical R and T states of the Monod-Wyman-Changeux model. In fact *all* of what is known about the allosteric behavior of ATCase can be accounted for not only qualitatively but *quantitatively*[16] by this model.

It remains to be seen how universally the Monod-Wyman-Changeux model will serve as an interpretation in general for the behavior of regulatory enzymes; for the moment it looks very attractive indeed.

References

1. E. R. Stadtman, *Advances in Enzymology*, Vol. 28, F. F. Nord, (Ed.), Interscience, New York, 1966, p. 41.
2. J. P. Changeux, *Scientific American*, April 1965, 36.
3. W. K. Maas and E. McFall, *Ann. Rev. Microbiol.*, **18**, 95 (1964).
4. H. S. Vogel, *Control Mechanisms in Cellular Processes*, D. M. Bonner (ed.), Ronald, New York, 1961, p. 23.
5. F. Jacob and J. Monod, *J. Molec. Biol.*, **3**, 318 (1961).
6. T. W. Rall and E. W. Sutherland, *Cold Spring Harbor Symp. Quant. Biol.*, **26**, 347 (1961).
7. J. Monod, J. P. Changeux, and F. Jacob, *J. Molec. Biol.*, **6**, 306 (1963).
8. J. Monod, J. W. Wyman, and J. P. Changeux, *J. Molec. Biol.*, **12**, 88 (1965).
9. J. R. Sweeny and J. R. Fisher, *Biochemistry*, **7**, 561 (1968).
10. D. E. Koshland, Jr., G. Nenethy, and D. Filmer, *Biochemistry*, **5**, 365 (1966).
11. R. A. Yates and A. B. Pardee, *J. Biol. Chem.*, **221**, 757 (1956).
12. J. C. Gerhart and A. B. Pardee, *J. Biol. Chem.*, **237**, 891 (1962).
13. J. P. Changeux, J. C. Gerhart, and H. K. Schachman, *Biochemistry*, **7**, 531 (1968).
14. J. C. Gerhart and H. K. Schachman, *Biochemistry*, **4**, 1054 (1965).
15. J. C. Gerhart and H. K. Schachman, *Biochemistry*, **7**, 538 (1968).
16. J. P. Changeux and M M. Rubin, *Biochemistry*, **7**, 553 (1968).

APPENDIX

Enzyme Classification

In 1930 there were approximately 80 known enzymes. In 1960 there were over 1300 known enzymes, and the number is continually expanded as modern research discovers more enzymes. Recognizing the growing problems of adequately naming and referring to these enzymes and the confusion that would arise in the absence of a systematic approach to enzyme nomenclature, an international commission on enzymes was set up by the International Union of Biochemistry in 1955 to consider, among other items, the classification and nomenclature of enzymes and coenzymes. Their work, led to the publication in 1961 of Volume 20 of the International Union of Biochemistry Symposium Series entitled *Report of the Commission on Enzymes*, wherein the presently used method of systematic enzyme nomenclature was proposed. This system has been subsequently adopted and is in universal use today.

The systematic nomenclature is to be used for all *single enzymes* and not applied to systems containing more than one enzyme. To name such multienzyme systems for the particular overall reaction that occurs, the word "system" should be used in the name. Thus the system catalyzing the oxidation of succinate by oxygen (succinate dehydrogenase, cytochrome oxidase and intermediate carriers) can be called the "succinate oxidase system" but not "succinate oxidase." The basis for naming the individual enzymes rests upon the grouping of enzymes according to the type of reaction catalyzed and the name(s) of the substrate(s). If some ambiguity exists due to alternative naming possibilities, the name of a prosthetic group (if the enzyme utilizes one) may be included to allow a unique naming. This should be avoided if possible.

Scheme of Classification and Numbering of Enzymes[1]

The commission has devised a system that provides both a classification of enzymes and a basis for numbering them.

Each enzyme number contains four elements, separated by points and arranged on the following principles:

1. The first figure shows to which of the six main divisions of the enzyme list the particular enzyme belongs. Enzymes can be divided into six main groups: (1) oxidoreductases, (2) transferases, (3) hydrolases, (4) lyases, (5) isomerases, and (6) ligases (synthetases). Lyases are enzymes that remove groups from their substrates (not by hydrolysis), leaving double bonds or, conversely, that add groups to double bonds; ligases, also known as synthetases, are enzymes that catalyze the joining together of two molecules coupled with the breakdown of a pyrophosphate bond in ATP or a similar triphosphate.

2. The second figure indicates the subclass. For the oxidoreductases it shows the type of group in the donors which undergoes oxidation (1 denoting a —CHOH— group, 2 an aldehyde or keto-group, etc., as shown below): for the transferase it indicates the nature of the group which is transferred; for the hydrolases it shows the type of bond hydrolysed; for the lyases it denotes the type of link that is broken between the group removed and the remainder; for the isomerases it indicates the type of isomerization involved; for the ligases it shows the type of bond formed.

3. The third figure indicates the sub-subclass. For the oxidoreductases it shows for each type of donor the type of acceptor involved [1 denoting a coenzyme (NAD$^{\oplus}$ or NADP$^{\oplus}$), 2 a cytochrome, 3 molecular O_2, etc.]; thus the first three figures of the number indicate clearly the nature of the enzyme, for example 1.2.3. denotes an oxidoreductase with an aldehyde as donor and O_2 as acceptor. For the transferases the third figure subdivides the types of group transferred (indicating whether the one-carbon group is a methyl or a carboxyl group, etc.), for the phosphotransferases, however, it is used to denote the type of acceptor as with the oxidoreductases. For the hydrolases this figure shows more precisely the type of bond hydrolyzed, and for the lyases the nature of the group removed. For the isomerases it shows more precisely the nature of the transformation, and for the ligases the nature of the substance formed.

4. The fourth figure is the serial number of the enzyme in its sub-subclass.

In a few cases it has been necessary to use the word "other" in the description of subclasses and sub-subclasses. They have been numbered 99, in order to leave space for new subdivisions, and this should be used to provide for similar cases which may arise in the future.

This method of numbering, in addition to its advantages as a system of classification and as an indication of the nature of the reaction catalyzed, avoids the chief disadvantage of consecutive numbering through the whole enzyme list, that the discovery of a new enzyme belonging to a class early in the list disturbs the numbers of all those following. By using the system proposed, it is possible to insert a new enzyme at the end of its sub-subclass without disturbing any other numbers. It is true that it may not be possible to place it immediately next to the enzyme it most nearly resembles, but it will at any rate be in the right group. Similarly, should it become necessary to create new classes, subclasses, or sub-subclasses, they can be added without disturbing those already defined.

The list of enzymes will certainly need revisions in the future, but it is intended that the numbers should remain permanently attached to the same enzymes as a definite means of identification. When the list is revised, the existing enzymes should not be renumbered to take account of any newly discovered enzymes; such enzymes should be placed at the end of their respective sections and should be given new numbers. This is regarded as very important. It is also important that new numbers should be allotted by authority (e.g., by future Enzyme Commissions or by a Standing Committee of the International Union of Biochemistry) and not by private individuals.

Systematic and Trivial Nomenclatures

The commission has given much thought to the question of a systematic and logical nomenclature for enzymes. There are two main difficulties. The first is the difficulty of obtaining a nomenclature that is really systematic without making wholesale changes in the existing nomenclature, which the commission was anxious to avoid as far as possible. The second is due to the fact that the systematic names of enzymes must include the names of their substrates, and since many substrates are complex substances, with long chemical names, it is inevitable that many of the systematic names will be too long for ordinary use.

It has been agreed that the best solution is that which is adopted by many other branches of science, including botany and zoology—to have two kinds of names. Accordingly the commission recommended two nomenclatures for enzymes, one systematic and one working or trivial. The systematic name of an enzyme is formed in accordance with definite rules, identifies the enzyme precisely, and shows the action of the enzyme as exactly as possible. The trivial name is sufficiently short for general use but not necessarily very exact or systematic; in a great many cases it is the name already in current use.

This system provides the advantages of a systematic nomenclature for the precise identification and classification of enzymes, while at the same time avoiding the confusion that would be caused by a wholesale renaming and retaining the advantages of conveniently short names for ordinary use. The greater part of the existing nomenclature will be retained as the trivial nomenclature; in a few cases, where there are good reasons, some changes will be made, but the commission is anxious not to make alterations unless they effect a worthwhile improvement.

The number might be thought to be a sufficient means of identification, but is useful for this purpose only when a copy of the list is available, and it cannot be used for enzymes not yet in the list. The systematic name by itself identifies the enzyme, and such names can be formed for new enzymes by the application of the rules, which give names that fit into the general pattern.

Where an enzyme is the main subject of a paper or abstract, it is recommended that its code number, systematic name (where a satisfactory one exists), and source should be given at its first mention; thereafter the trivial name may be used. Enzymes that are not the main subject of the paper or abstract should be identified at their first mention by their code numbers. When the paper deals with an enzyme not yet in the Enzyme Commission's list, the author may introduce a new systematic name and/or a new trivial name, both formed only according to the recommended rules, but a number should be assigned only by a standing committee of the International Union of Biochemistry, or by a new Enzyme Commission.

It is particularly important that the systematic nomenclature be employed in abstracts. In such brief publications the possibility of mistaking the identity of an enzyme is highest and has most serious consequences. Therefore it is recommended that in all abstracts of papers in which the enzyme is the main subject, the code number, systematic name, and source be given for the fullest description of the enzyme.

Key to Numbering and Classification of Enzymes[2]

1 *Oxidoreductases*

   1.1 Acting on the CH—OH group of donors
      1.1.1 With $NAD^{\oplus}$ or NADP as acceptor
      1.1.2 With a cytochrome as an acceptor
      1.1.3 With O_2 as acceptor
      1.1.99 With other acceptors
   1.2 Acting on the aldehyde or keto group of donors
      1.2.1 With $NAD^{\oplus}$ or NADP as acceptor
      1.2.2 With a cytochrome as an acceptor
      1.2.3 With O_2 as acceptor
      1.2.4 With lipoate as acceptor
      1.2.99 With other acceptors
   1.3 Acting on the CH—CH group of donors
      1.3.1 With $NAD^{\oplus}$ or NADP as acceptor

3.4 Acting on peptide bonds (peptide hydrolases)
    3.4.1 α-Amino-acyl-peptide hydrolases
    3.4.2 Peptidyl-amino acid hydrolases
    3.4.3 Dipeptide hydrolases
    3.4.4 Peptidyl-peptide hydrolases
3.5 Acting on C—N bonds other than peptide bonds
    3.5.1 In linear amides
    3.5.2 In cyclic amides
    3.5.3 In linear amidines
    3.5.4 In cyclic amidines
    3.5.5 In cyanides
    3.5.99 In other compounds
3.6 Acting on acid-anhydride bonds
    3.6.1 In phosphoryl-containing anhydrides
3.7 Acting on C—C bonds
    3.7.1 In ketonic substances
3.8 Acting on halide bonds
    3.8.1 In C-halide compounds
    3.8.2 In P-halide compounds
3.9 Acting in P—N bonds

4 *Lyases*

4.1 Carbon-carbon lyases
    4.1.1 Carboxy lyases
    4.1.2 Aldehyde lyases
    4.1.3 Ketoacid lyases
4.2 Carbon-oxygen lyases
    4.2.1 Hydrolyases
    4.2.99 Other carbon-oxygen lyases
4.3 Carbon-nitrogen lyases
    4.3.1 Ammonia lyases
    4.3.2 Amidine lyases
4.4 Carbon-sulfur lyases
4.5 Carbon-halide lyases
4.99 Other lyases

5 *Isomerases*

5.1 Racemases and epimerases
    5.1.1 Acting on amino acids and derivatives
    5.1.2 Acting on hydroxyacids and derivatives
    5.1.3 Acting on carbohydrates and derivatives
    5.1.99 Acting on other compounds

References

1. *Report of the Commission on Enzymes*, IUB Symposium Series, Vol. 20, Pergamon Press, New York, 1961. Most of this appendix is adapted directly from this reference.

2. T. E. Barman, *Enzyme Handbook*, Vol. I, Springer Verlag, Berlin, 1969, p. 16.

Index